THE STONES OF BRITAIN

Also by Richard Muir

MODERN POLITICAL GEOGRAPHY

THE ENGLISH VILLAGE

RIDDLES IN THE LANDSCAPE

GEOGRAPHY, POLITICS & BEHAVIOUR (with R. Paddison)

SHELL GUIDE TO READING THE LANDSCAPE

THE LOST VILLAGES OF BRITAIN

TRAVELLER'S HISTORY OF BRITAIN AND IRELAND

HISTORY FROM THE AIR

NATIONAL TRUST GUIDE TO PREHISTORIC & ROMAN BRITAIN (with H. Welfare)

VISIONS OF THE PAST (with C. C. Taylor)

EAST ANGLIAN LANDSCAPES, PAST AND PRESENT (with Jack Ravensdale)

THE SHELL COUNTRYSIDE BOOK (with Eric Duffey)

SHELL GUIDE TO READING THE CELTIC LANDSCAPES

THE STONES OF BRITAIN

Richard Muir

Michael Joseph
London

First published in Great Britain by Michael Joseph Ltd
44 Bedford Square, London WC1
1986

© 1986 by Richard Muir

All Rights Reserved. No part of this publication
may be reproduced, stored in a retrieval system,
or transmitted in any form or by any means, electronic,
mechanical, photocopying, recording or otherwise,
without the prior permission of the Copyright owner

British Library Cataloguing in Publication Data

Muir, Richard, *1943–*
The stones of Britain.
1. Stone——Great Britain——History
I. Title
620.1′32′0941 TA426

ISBN 0 7181 2539 8

The endpaper map showing the geological
structure of the British Isles is
© George Philip & Son Limited and is reproduced
with their permission.

Typeset in Photina
Filmset and printed by BAS Printers Limited,
Over Wallop, Hampshire

Contents

Introduction

Stone is something that we tend to take for granted. We certainly appreciate the tones and textures of different rocks when they are displayed in the walls of churches, cathedrals or cottages, and we may travel miles to admire the boulder architecture of a prehistoric tomb or circle. Guide books to castles, stately homes and ancient monuments abound, yet it is very seldom that we are told much, or anything, about the origins of the building materials, how they were obtained, and the stone crafts involved in assembling them.

This book is a small attempt to redress the balance. The topic of stone and its use by man is a vast one and I may only have scratched its surface; but I will be well content if I can encourage the reader to look a little more closely at the natural and man-made outcrops of stone in the landscape. In Britain and Ireland, the legacy of monuments is enormous. It goes without saying that looking at the innumerable facets of the man-made and natural scene is a source of pleasure. Looking, thinking and asking questions converts this pleasure into a stimulating challenge, often an absorbing hobby, and always a delight. A medieval cathedral, ruined abbey or castle almost always has a considerable aesthetic appeal, but it becomes far more interesting when we puzzle about where its building blocks came from, how they were conveyed from the quarry to the building site, what the stones cost, and what sort of men commissioned the works, supervised the masons, and dressed or carved the stones.

I have organized the book into three parts. The first is concerned with stone in the landscape. Here I describe a series of distinctive 'total landscapes', and draw attention to the relationship between the natural underlying stones of different regions and the use of their assets by man. The second part covers the remarkable uses of stone by prehistoric communities – uses which were so fundamental to the economic and spiritual lives of the people concerned that without stone the progress towards civilization might never have been accomplished. The third part relates to the uses of stone during the historical period, and describes the rise and fall of masonry and quarrying through seasonal metaphors. Sadly, of course, we are living in the deep mid-winter of the stone building era, and most skilful applications of masonry now are concerned with restorations.

I have not attempted to write a geology book; for all its very real interest, geology is a technical subject dealing with advanced concepts of chemistry and physics. Indeed, in seeking to pass on to the general reader an enthusiasm for rocks and tools, buildings, walls and monuments made of stone, I have sought to reduce the technical terms to an absolute minimum, hoping to write simply, though not superficially. In fact, this is as much a book about people as about stone. Good and bad landscapes, just like good and bad buildings, are essentially man-made, and therein lies their basic interest and appeal.

Every day during the holiday season, thousands of people leave bad, new or nondescript buildings to look at others which are old, finely wrought and textured, and a delight to behold. Despite the existence of this widespread fascination, little is taught in the classroom which assists the understanding and appreciation of our endowment of monuments. My work takes me to scores of churches, castles and prehistoric constructions, and I see hosts of people who are no less enthralled than myself by the things that are displayed. This interest in the past and the questions the visitors ask testify to an alertness which is reassuring when we consider the glaring gaps and blunders in the fields of state education and conservation. In its limited way, I hope that this book may make some of the questions more pertinent, and suggest others which might have been overlooked.

PART ONE

STONE IN THE LANDSCAPE

Introduction

The British Isles can be likened to a mosaic which is composed of many different and distinctive landscapes. This wonderful diversity is, to a very considerable extent, a reflection of the remarkable variation in the geological roots of our landscape. One does not need to be a geologist or a geographer to appreciate that the scenery of the South Downs or Dartmoor contrasts in the most fundamental ways with that of (say) Orkney or the Lake District. It is no less apparent that many of the scenic differences and delights which we enjoy are imparted to the landscapes by the varying natures of their native rocks. These first chapters are concerned with 'total landscapes', in which the works of nature and of man are married. In any appreciation of landscape, one cannot separate geology from history. Wherever we look, we will find that the rocks which underlie or outcrop in the natural scene have greatly influenced farming practices, have often allowed the development of local industries, and have guided the course of ecclesiastical and vernacular building styles. Perhaps more than any other factor, the native rocks provide the key to an understanding of the essential characters of places in the British Isles.

Such is the diversity of the landscape jigsaw of the British Isles that it is impossible here to detail every component. Instead, I will offer some broad remarks concerning the character of the natural and man-made scenery which has developed upon the surface of the main rock types, and highlight certain landscapes whose visible charms are determined by the rocks beneath. Whether one is visiting places where the geological roots of scenery assert themselves in a thunderously insistent manner, such as the Lake District or the Burren, or places where rock presents itself in a soft-spoken and subtle manner, such as parts of the English Midlands, an appreciation of the relationship between stones and man can only intensify the pleasures of touring and rambling in Britain.

Chapter 1

The Hard Rock Landscapes

*H*ard rocks – granites, basalts and their relatives, and most changed, or 'metamorphic', rocks – often give rise to scenery which is as violently dramatic as the forces which created them. Such rocks are often associated with rugged upland or mountain scenery and, although many qualifications to this statement might be made, on the whole it is because of their hardness that these rocks tower above adjacent landscapes. The hard rocks have a variety of origins: some, like the granites, are the products of the slow cooling of molten rock bodies deep beneath the surface of the earth; the basalts, in contrast, often gushed forth from surface or submarine fissures to harden swiftly as level sheets; the tough slates resulted from the baking of softer rocks in the terrestrial pressure cooker. Their toughness and contrasting origins apart, the hard rocks are a very mixed bunch in terms of their constituents and structure, the types of scenery which they weather to produce, and their uses to man. The range includes the moor and tor scenery of Dartmoor, the slate and volcanic rock mountain landscapes of the Lake District and North Wales, the bizarre local basalt scenery of the Antrim coast and Staffa Island, and many others. Occasionally, as in parts of Ulster, the scorched or fire-born rocks can form scenery which is low-lying or rolling rather than rugged. Generally, however, the hard rocks produce tough, uncompromising landscapes, thinly peopled by hill-farming families as hardy as the rocks themselves.

Although it is not a particularly common rock, wherever in Britain granite does outcrop at the surface, it tends to create its own distinctive scenery. The origins of granite remain controversial. Some geologists hold that granite forms directly from the setting and crystallization of a hot magma of molten rock. Others believe that it forms from existing rocks coming into contact with a fiery body of magma, which has caused them to melt and recrystallize. In any event, granite forms deep beneath the surface of the

The beautiful Cuillin Hills of Skye are formed of gabbro, a rock with origins rather like those of granite, but with a different composition of minerals. Vigorous glaciers have helped to sculpture the scenery.

earth, and so it can only be seen as a surface outcrop when erosion has stripped away great thicknesses of overlying rocks. These deeply subterranean origins are apparent in the large sizes of the crystals which result from the gradual cooling of the water-rich magma. Plainly seen in any granite boulder, building stone or tombstone are the translucent, whitish crystals of quartz, grey or pink feldspars, and the flake-like crystals of sparkling mica, which may be black or silvery white.

Granite is, under normal circumstances, a very hard rock. Thus, once erosion has stripped away the overburden of other rocks, it will tend to erode more slowly than its surrounding areas, and will form mountain domes or swelling uplands. Much of Cornwall is underlain by a great granite mass, or 'batholith', which appears at the surface to form a series of bosses: Dartmoor, Bodmin Moor, Land's End, the Scilly Isles, and others in the region of St Denis and Mawgan. In northern Scotland, numerous granite mountains in the Grampians and Cairngorms consist of masses which surface like schools of whales above the storm seas of ancient shists and gneisses that surround them. Powerful granite mountain scenery can also be seen in widely separated parts of Ireland – the Wicklow Mountains to the south of Dublin, the desolate ranges in the northern flank of Galway Bay, the Glendowan Mountains of County Donegal, and the Mountains of Mourne in County Down. Elsewhere, granite may surface to form more solitary sum-

Roche Rock in Cornwall is a dramatic granite outcrop.

mits, like the bleak, peat-carpeted dome of Cheviot, which swells above the surrounding volcanic and limestone rocks of Northumberland to a height of 2,676 feet, or the granite island summits of Galloway in the Southern Uplands.

Despite its rugged constitution, granite is vulnerable to the processes of chemical decay which are mainly associated with the warmer climatic regions. As we know only too well, the present British climate is not subtropical, nor even warm temperate, but the climate was much warmer in the long Tertiary period which preceded the Ice Ages of the Quaternary era. In the course of cooling from a molten magma, granite masses tended to contract and become fissured by horizontal joints. Under warm conditions, the chemical agents of decay will exploit these fissures in granite masses which lie at or near the surface, attacking the feldspar constituents, and thus undermining the structure of the crystalline rock.

Meanwhile, the surface of the granite may be exposed to wide diurnal ranges of temperature, and differences in the rates at which the different crystalline contents of the rock expand and contract produce stresses which also lead to disintegration. The effect of these forms of weathering is to produce a landscape of smooth-surfaced, dome-shaped mountain masses, with slopes blanketed in rotten rock debris and littered with the large, detached stone blocks which have been dislodged from the bedrock through chemical

Haytor Rock has a smoother profile than some other granite tors on Dartmoor; the figures on the summit provide a scale.

attack along the joints. Although the granite scenery of the Cairngorms has been changed in detail by the effects of glaciation, its smooth slopes and rounded summits are essentially those of a sub-tropical granite landscape, and this view is supported by the great thicknesses of chemically-rotted rock which swathe many slopes.

Although one would not describe the granite scenery of Dartmoor and the Cornish moors as either pretty or breathtaking, it is, in a bleak and soulful way, dramatic. Certainly it embraces a series of landscapes which are packed with interest, whether one is a geologist, a prehistorian, or simply a lover of fascinating places. The granite bosses of Dartmoor and Bodmin Moor rise in smoothly-swelling blisters, often culminating in rocky 'tors' whose knobbly silhouettes contrast with the sweeping undulations of the surrounding moor. The origins of the tors have been much debated, but it seems that they were created by two distinct climatic regimes. Firstly, chemical weathering worked along the fissures in the subsurface rock to widen the joints and rot great masses of granite. Then, perhaps in the bitter periglacial conditions experienced when much of the land to the north was under ice, the decayed rocks were removed, leaving tottering piles of unrotted slabs and boulders as hilltop landmarks. Tors like the Cheesewring and Brown Willy on Bodmin Moor, and East Mill Tor and Haytor Rocks on Dartmoor are good examples, with granite slabs weirdly stacked like loaves of unleavened bread in a bakery.

In the area of St Austell, nature and man have combined to produce scenery which is even more outlandish – the ghostly, cone-shaped mountains which are the refuse tips of the china-clay industry. As granite cooled from a water-rich magma, scorching bodies of water and steam were released through the higher domes of the magma intrusions, destroying the micas and the giant feldspar crystals which had grown there. Kaolinite, or china clay, resulted from the destruction, which left only the quartz crystals intact. With the rise of the Staffordshire pottery industry, vast quantities of china clay were shipped out from Cornish ports; and the white mountains of southern Cornwall consist of the quartz grains which remained after the kaolinite had been flushed from the rotten granite.

Before geology provided a basis for the eighteenth- and nineteenth-century tin- and china-clay-mining booms, Cornwall was poor. It still is, so far as its indigenous people are concerned. The long-standing agricultural poverty which has caused so many native families to seek their fortunes from the sea, mines or quarries is largely the fault of geology. The moors are sufficiently high to produce a problem of exposure, which is exacerbated by the storms that roll in unimpeded from the Atlantic; but its slopes are insufficiently steep to achieve a swift evacuation of the heavy rainfall. Rather than thirstily accepting water like a limestone, the granite tends to hold it in treacherous, peaty bogs, while the rock erodes to produce a coarse, sandy, acidic soil.

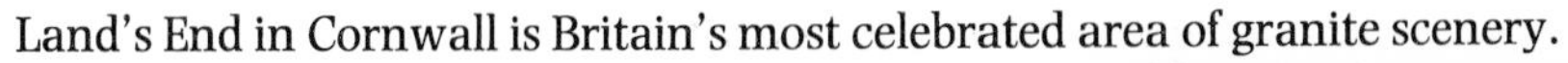
Land's End in Cornwall is Britain's most celebrated area of granite scenery.

China clay workings near St Austell in Cornwall.

Peat moor and bog carpet most of the granite uplands, and we cannot be certain to what extent this impoverishment results from climatic deterioration which began at the end of the Bronze Age, or from prehistoric man's removal of the natural deciduous forest and the overgrazing of the moors. Both must have played a part. Even so, the granite moors are liberally sprinkled with monuments to prehistoric life – stone circles like The Hurlers, portal dolmen tombs like Trethevy Quoit, and villages like Grimspound – and Dartmoor is littered with the ruins of Bronze- and Iron-Age huts. The ancient field walls, or 'reaves', and field systems which they outline occur in several places on the moors, and may bear witness to a richer environment in times gone by, but it is hard to imagine that this was ever prime farming land. Rather, the remains remind us of the wealth of prehistoric relics which must have been lost from lower, more favoured areas as a result of the continuing success of arable farming.

What the moors did offer was moorstone: granite building blocks detached from the bedrock by chemical attacks along the joints of a parent mass, which could be dug from the ground or gathered from the rock debris, or 'clitter', of a ruined tor. Moorstone was used in the building of rugged and unsophisticated homes and churches. Neat ashlar work and fancy carving were precluded by the toughness of the granite, and the ability to carve and shape it does not seem to have been acquired until the medieval period

Trethevy Quoit in Cornwall, the burial chamber of a prehistoric tomb built of massive slabs of moorstone.

had almost run its course. Professor W. G. Hoskins, the great economic and landscape historian, mentions the granite church tower at Probus, which dates from the 1520s, as a very early example of the use of cut granite stones.*

As in so many other hard-rock areas, which are peopled on the whole by thrifty, impecunious hill farmers, the homes and churches of Cornwall are simple, spartan buildings which display the local stones in a way that harmonizes with their bleakly romantic settings. Buildings of moorstone abound in Cornwall, but granite is not the only local stone on display. Many of the older cottage roofs use Cornish slate, and it was quite common for a rainproof cladding of slates to be hung on the weather side of a house.

*_One Man's England_ (BBC, 1978)

Cornwall did not escape the nineteenth-century advance of Welsh slate, but since the principality has for long been largely devoid of timber, lacks good brick-clay deposits, and produces its own local slates, it can still boast a wealth of vernacular buildings in the native stone. (It also possesses a large number of tatty and nondescript holiday homes which greatly detract from the quality of its landscape.)

The largest of the Cornish slate quarries is the Delabole, which has been worked since at least the closing years of the medieval period and has reached depths to compare with those of the Aberdeenshire granite quarries at Rubislaw and Kemnay (with which it competed for the title of Europe's deepest quarry). Blue-grey Delabole slate still crowns many Cornish houses. During the last century, around a thousand men worked at the quarry, and its products were shipped out of the tiny port of Port William to many distant destinations.

Two other Cornish stones deserve a mention. The Serpentine rock of the Lizard peninsula contains a little feldspar, but consists mainly of a green mineral. It was once widely favoured as a tough, decorative stone which could be polished to display unusual warm green or red mottled hues. Near

The gigantic crater of the Delabole slate quarry in Cornwall; an idea of the scale of the workings can be gained from the heavy vehicles that are turning the first corner on the right on the quarry road.

its source, Serpentine was also used as a rough, dull green stone in the cottages of the Lizard area, or occassionally combined with pale granite in the building of two-toned churches. The rock's main claim to fame now derives from the remarkable stability of the Serpentine mass, which provides a firm platform for the radio dishes of the Goonhilly Earth Station. Polyphant is the name of a little Cornish village, but it is also (as addicts of TV and radio quiz games will know) the name of a stone. The small and long-abandoned quarry at Polyphant produced a blue-grey product, which could not only be used for structural work in church building and for tombstones, but could also be polished as a soft-tinted, decorative stone.

There are many parts of the British Isles – like Northamptonshire, most of Somerset, and County Meath – where stone provides the fine detail in the landscape while not asserting itself boldly in the natural scene. There are others – and Cornwall is one – where stone is not only apparent in the tracery of the field walls, cottages and churches which stud the scenery, but is also clearly the substance of the landscape itself. As is so often the case, one cannot improve upon the prose of W. G. Hoskins. Asking what it is that makes Cornwall so distinctive, he suggests that '... more than any other part of England, Cornwall is a land of stone. Everything in it, everything of any age at all, is made of stone. The old buildings came out of the ground under men's feet ...'*

Although it neither provides the basis for much spectacular mountain scenery nor features prominently in the stone buildings of Britain, basalt must be mentioned at least briefly. Like granite, basalt is a fire-born or 'igneous' rock, but it differs from granite in several important ways. Instead of having crystallized slowly, deep beneath the surface of the earth, basalt is a volcanic rock which has blasted a passage through the crust to spill out in great sheets of molten rock upon the surface or seabed. The large crystals which we associate with granite have not had time to form, since the basalt lavas cooled very rapidly in contact with air or water. Basalt is a dark grey rock with the texture of fine sandpaper and its tiny constituent crystals of feldspar, augite, and olivine can scarcely be identified with the naked eye. Being tough, localized in its occurrence, and rather dull, basalt has had little appeal for the mason; but, very occasionally, one may see the stone polished in decorative work, and in this state it becomes quite agreeable.

Although it is seldom associated with magnificent scenery, in at least two parts of the British Isles, basalt has produced uniquely dramatic effects. In the celebrated Giant's Causeway of County Antrim, and at Fingal's Cave on the remote Scottish island of Staffa, the rock can be seen forming weird

Ibid

Basalt flows form the distinctive cliffs around Kilt Rock on Skye.

Erosion of the basalt rocks has produced the remarkable pinnacle of the Old Man of Storr on Skye.

tableaux of closely-packed columns. Most of the columns are six-sided, though the range is from triangular to octagonal sections. The whole effect, which resembles clustered organ pipes of different heights, results from the contraction of the basalt in the course of its rapid cooling.

Elsewhere, near-horizontal sheets of basalt may be seen, and these too are ancient lava flows. Lough Neagh in County Antrim lies in the sunken centre of Britain's largest lava plateau, while in several parts of the county, columnar structures can be seen in the faces of dark basalt cliffs at the eroded margins of great lava sheets. Submarine eruptions sometimes coated the surrounding seabed with a thick icing of basalt, before a return to more stable conditions allowed the deposition of beds of new limestone. Alternating conditions of basalt and limestone formation are evident in some Derbyshire landscapes, where the rock faces are banded by thin sheets of dark basalt, which contrast with the ghostly hues of the limestone. Knolls which can be seen rising above the plateaus to the west of Matlock and near Castleton are the remains of the pipes from which lava erupted before their vents became choked with volcanic debris. The Sidlaw Hills, lying between Dundee and Perth, clearly display the layered accumulations of ancient lava flows, and stand like plywood rafts above the surrounding farmlands.

We tend to regard the English Midlands as an area of smoothly undulating agricultural countryside, criss-crossed by hedgerows and dappled by the patterns of fat lowland villages and industrial or market towns. The best of the scenery is mellow and inviting, and we do not expect to encounter the rugged dramas of hard-rock scenery. Even so, just to the north-west of Leicester and deep in the heart of the Midlands, the rocky island of the Charnwood Forest forms a starkly attractive contrast to the surrounding lowland patchwork of hedged fields. The forest is a strange geological exclave, where the ancient basement of pre-Cambrian rocks surfaces through the blanket of younger rocks and becomes visible in the jagged crags which crown many of the Charnwood hilltops. The complex crystalline rocks include a variety of metamorphic and volcanic types, and also granites, which have been quarried for roadstone at Mountsorrel, while the slates of Swithland, which have been exploited for use in the local roofs, resemble those of Wales. Very gradually, the erosion of the surrounding red sandstone marls is uncovering a landscape which is older than life itself. The marl still fills the valleys of lifeless rivers, which were engraved into the old rock surface hundreds of millions of years before fish evolved on earth. Although the tallest Charnwood summits are only around nine hundred feet in height, the forest, with its angular, ice-shattered peaks, is like a fragment from the western mountains incongruously lodged in the gentle countrysides of the Midland Plain.

A few of the most memorable British landscapes are engraved in a single, distinctive type of rock – Dartmoor is an example. Others derive much of their appeal from the combination of two or more rock 'personalities'. The Highlands of Scotland, for example, are cut across a wide variety of types of ancient metamorphic rocks, with the frequent exposures of granite adding to the diversity of the scene. The landscapes of Snowdonia and the Lake District are similarly varied, their metamorphic rocks, such as slates, being juxtaposed with other rocks of volcanic origin. Metamorphic rocks such as schists, slates and gneisses produce some of the finest mountain scenery in Britain, including the mountains of Counties Donegal and Connemara, and superb chains of sea-cliffs, like those to the north of Aberdeen. Almost invariably, the geological roots of such enchanting scenery are too complex to be briefly described. Instead, just one region – the English Lake District – will be examined in detail.

In any broad discussion of scenery that is both beautiful and dramatic, England must yield to the superior claims of Ireland, Scotland, and Wales, for the Celtic lands have more uninterrupted expanses of breathtaking scenery. However, it can equally be argued that England contains the single most lovely landscape of all. Although the Lake District sometimes seems unable to digest its mounting numbers of visitors, it is hard to imagine that there

The landscape of the Scottish Highlands is carved in ancient 'metamorphic' rocks which are periodically breached by exposed granite masses. This photograph is of Loch Long and Ben Arthur,

is any other part of the world which can offer such a varied concentration of compelling views. There are, in fact, two Lake Districts. The one, which we savour from lake shores or glimpse on picture postcards, is mellow and pastel-tinted. The other, which shepherds, fell walkers and climbers know, is vast, brutal and humbling. Both these landscapes are hewn in rock. The ubiquitous stones form the dwellings, the curving tracery of field walls which partition the lower fells, the bijou bridges and sturdy little churches, and, much more rarely, the stone necklaces of prehistoric temples.

The story of man in the Lake District has been enacted in a landscape which, for all its visual rewards, is rugged and uncompromising. For most of the historical period, Lakeland was widely regarded as a wild and inaccessible backwater of England – and so it was. Scarcely more than two centuries ago, the traveller had difficulty in penetrating northwards beyond Kendal except on horseback. It was only in the 1790s that the first stalwart ramblers began to arrive, many risking the hazardous coach trip across the tidal sand-flats of Morecambe Bay. Poets, aesthetes and publicists of the region's beauties formed a vanguard. Coleridge was the first tourist to climb Skiddaw, but it fell to John Ruskin to ask: '... first what material there was to carve, and then what sort of chisels, and in what workman's hands, were used to produce this large piece of precious chasing, or embossed work, which we call Cumberland and Westmorland ...'

It is unlikely that Ruskin realized the complexity of the question he had posed. While most of the rocks which form the Lake District are more than 500 million years old, that part of the earth which is now occupied by the district existed variously as sea, swamp and mountain range during the intervening periods. At different times in the geological past, rocks were laid down, uplifted into mountain masses at least twice, then eroded away to expose the roots of older rocks, until the cycle of submergence, deposition, and mountain building began again.

More than twenty-five million years ago, the same stupendous earth movements which created the Alps caused the levelled, ancient rocks of the Lake District, as well as the thick deposits of younger sediments which blanketed them, to be thrust upwards to form a great dome. At once, the forces of erosion renewed their onslaught, as hurtling streams radiating outwards from the high domed core of the mountain mass cut through the younger rocks, stripping away the vast thicknesses of overlying strata to expose the ancient contorted roots of the slate and volcanic formations beneath.

Around two million years ago, the climate became progressively cooler. Glaciers began to form in the loftiest Lakeland valleys – slithering slowly down the valley channels, merging, spilling out beyond the mountains, and combining, in the Vale of Eden and Morecambe Bay, with great ice rivers creeping south from Scotland. Above the glaciers, stresses from the freezing and thawing of water droplets trapped in their crevices caused rock fragments to shatter from the cliffs and slopes. Falling into the glaciers beneath, these angular chunks of scree debris were welded in the ice to form abrasive teeth, which scoured the valley sides as the ice streams slowly advanced.

As Ice Ages came and went and glaciers repeatedly advanced and retreated, a distinctive, fiercely-glaciated landscape was created. Valley troughs were deepened, smoothed and straightened; at the heads of the valleys, basin-like corries – the lofty nurseries of the glaciers – were slowly scooped out; knife-sharp arretes were honed as the steep corrie backwalls retreated and met; hummocks of ground-up rock debris were then dumped as moraine when the glaciers began to recede; and in the aftermath of glaciation, lakes formed in the hollows which the ice had gouged in the valley bottoms. Gradually, the deposition of river silts is filling up the lake basins, and eventually the lakes will disappear.

If we could compress time in such a way that the entire process of glaciation took place yesterday, then most of the rocks into which the tortured

A Skiddaw Slate landscape seen from Cat Bells, which overlooks Derwentwater. Skiddaw mountain forms the skyline to the right and the market town of Keswick is near its foot.

glacial landscape was engraved would be a year or more in age. Three rock types in particular form the foundations of Lakeland scenery, and although each is ancient and has endured a similar traumatic history of uplift, exposure, and glacial torment, each has stamped its personality on the mountain scene.

To the north-west of a line which runs roughly from the south-west to the north-east a little to the south of Keswick are the Skiddaw Slates. Unlike those of the Snowdonian quarries, the Skiddaw Slates cannot be cleaved to form first-rate, slender roofing slates, and their shortcomings as exportable building materials have surely saved the Lakeland landscape from the worst ravages of quarrying. The sands, silts, and muds which eventually formed the Skiddaw Slates accumulated around 500 million years ago in a wide ocean trough. Their conversion into slates took place during the great 'Caledonian' period of mountain building, over 400 million years ago, while the sediments, which had accumulated to thicknesses of several thousands of feet, were again compressed and contorted in a second great eruption of mountain building around 300 million years ago. Dark and greenish in colour, the Skiddaw Slates are relatively homogenous in their resistance to erosion. Consequently, the scenery which they support reflects much more the processes of weathering – notably glaciation – than the effects of local differences in the toughness of the rocks. Skiddaw Slates scenery is organized around the theme of smoothly sweeping slopes; it is imposing, sometimes

The rugged scenery of the Borrowdale Volcanics in the Langdale Pikes area of the Lake District. Cold Pike is the central summit.

even majestic, but lacks the savage, precipitous, craggy details of the more dominating landscapes to the south. Even so, Skiddaw itself is no molehill, but a 3,000-footer, even if it does lack the ominous presence of some of the more southerly giants, like Scafell Pike, Pillar, or Great Gable.

Forming a broad band to the south of the Skiddaw-Slate country, and bounded to the south-east by a line which runs roughly from south-west to north-east, almost through Ambleside, are the rocks of the Borrowdale Volcanic Series. As the ocean trough in which the sands, silts, and muds which formed the Skiddaw Slates slowly filled, a frenzy of volcanic activity began. The ocean sediments were bombarded by the hurtling debris of frequent eruptions, blanketed in outpourings of lava, and smothered by volcanic ash, as the vents of volcanoes rose above the ocean in chains of fiery pustules spewing bombs, lava and ash across the surrounding seas. The volcanoes, in turn, were eroded and the chaotic residue of muds, 'tuffs' (composed of small ash fragments), 'volcanic breccias' (formed from larger, angular fragments) and lava flows were compressed and contorted by the same mountain-building traumas which converted the Skiddaw Slates, so that they often emerged as green, slate-like rocks.

The mountain scenery of the Borrowdale Volcanics, however, is of quite a different nature. Being composed of a chaotic mixture of changed, or 'metamorphosed', sedimentary and volcanic debris, the Borrowdale Volcanics juxtapose rocks of greatly differing quality and hardness. Erosion has scoured and nibbled at the softer rocks, leaving the more resistant ones, notably the compressed lavas and the most intransigent tuffs, to jut forth as overbearing crags and ridges. This scenery, with its noble, vertiginous cliffs, can be breathtakingly beautiful. Seen from a distance on a fine day, it is as lovely as anything which the landscapes of Britain can offer. But on those not-infrequent days when the skies are leaden and the cloud veils shift and swirl amongst the crinkled crags, even accessible places like the Honister Pass call to mind 'The Valley of the Shadow of Death'.

Moving southwards from the fierce splendour of the Borrowdale Volcanics landscapes – and leaving behind the great peaks like Helvellyn, Bow Fell, and Scafell, as well as the awesome crags like Gimmer, Long Stile, and Raven which stud and scar it – we enter the third and southernmost component of Lakeland geology, the Upper Slates. These are a complex series of sedimentary rocks more than 400 million years old, including shales, sandstones, limestones and mudstones – some of which have been converted into slates and resemble those of the Skiddaw area. Approaching the Upper Slates from the north, the traveller first encounters the narrow band of Coniston Limestone, which runs across the northern tip of Lake Windermere and succours a long and slender verdant strip of calcium-loving plants that flourish between the sour, acid fastnesses on either side.

I find the scenery of the Upper Slates rather too sweet and gentle to be regarded as a full partner in the Lake District proper: it is a charming and lushly-appointed foyer which the southerner moves through to reach the sterner drama of the theatre itself. The Upper Slates now form only the southern flanks of the great dome of the Lake District, erosion having stripped them from its crown to expose the sterner stuff of the Borrowdale Volcanics which lie beneath. They provide more detailed scenery than the Skiddaw Slates because of their varied layering of slate, grit, limestone, and shale; and in places, the harder components, like the grits, form faint escarpments. Subdued, in comparison with the violent landscapes to the north, hardly any of the summits in the area exceed 1,000 feet and most are in the 800-foot range. The landscape is thus knobbly rather than mountainous and, with better soils than those further north to support deciduous woodland and greener pastures, idyllic rather than imposing.

These, then, are the three main components of the Lakeland scene, though of course the detailed picture is still more complicated. The younger but still very old rocks which probably once blanketed the great mountain dome have been eroded away to expose the more ancient rocks, and the Carboniferous limestone thus survives only as a great girdle, ringing the District on all quarters but the south-west. Its inward-facing scarps form a stockade around Lakeland, and much greener pastures are seen growing on its limey soils.

At the time of the volcanic traumas which created the Caledonian mountain ranges, vast oceans of molten rock burst through the sediments that lay around the roots of the upswelling mountains, thousands of feet below the surface. After cooling slowly to form granites, these rocks were exposed by the erosion of the overlying strata hundreds of millions of years later. Lakeland granite scenery is less compelling than that of some other places, such as the Cairngorms, though rocks of the granite family form the pinkish summits of Red Pike and Caw Fell, and also occur in two great pockets – one around the eastern half of Ennerdale Water, the other in Eskdale, on the south-western flanks of the district. Syenite, a coarse-grained cousin of granite, forms the summit of High Stile, which faces Red Pike above the waters of Buttermere far below; and small pockets of granite-like rocks are found near the quarrying village of Threkeld, and near Skiddaw. Finally, the district is criss-crossed by 'dykes', formed when sheets of molten rock have been injected along lines of weakness, such as fault planes, to breach older rocks.

Naturally, the geological complexity of the Lake District is much greater than I have described, but any reader who is aware of the basic differences between the three main rock types – the Skiddaw Slates, the Borrowdale Volcanics, and the Upper Slates – will swiftly recognize their different contributions to the making of scenery. Fascinating as the relationship is, being a landscape historian rather than a geologist, I find the human contribution to the creation of our surroundings much more absorbing. Though wild, the Lake District that we see today is as much a product of human endeavour as are the Midland plains and southern downlands. Stone of one kind or another was always close to hand, and from the earliest times it was an essential adjunct to human life.

Relatively little is known of the Middle Stone Age bands of hunters and fishers who first penetrated the Lakeland passes. They entered a landscape being re-colonized by trees advancing northwards in the wake of the vanished ice sheets and dwindled glaciers. The Lakeland scenery then was very different from that which modern visitors relish; the upland slopes were clad in endless swathes of pine and birch, while oak, elm, alder, ash and hazel cloaked the lakeshores and lower slopes. The hunting folk wrought finely-chipped tool-kits and missile-heads from stone, but, before many millennia had passed, stone came to occupy an even more central role in Lakeland life, and to play a leading part in the creation of the scenery of sweeping grass- and bracken-clad fell slopes which we see today. The new Stone Age witnessed the remarkable transformations of life and landscape made possible by the introduction of farming, which gradually superseded the old hunting and gathering lifestyle. Some of the earliest evidence of the changes has come from the analysis of ancient pollen grains, moistly sealed and preserved in deposits beneath lakes, tarns and bogs. Around 4400 BC, it seems, trees such as the elm and oak began to retreat, while grasses and cereal crops replaced them over considerable areas.

Clearly, these events mark the arrival of pioneering farming communities, who were deliberately removing natural woodland to create arable plots and grazing ranges. In the re-fashioning of the Lakeland landscape, one tool was essential: the stone axe. In the hands of successive generations of Neolithic peasant farmers, the axes chipped and hacked, slowly reaping the wildwood. Once felled, the timber was probably burned to enrich the soils. On the upper fellsides, the relentlessly munching sheep and cattle ensured that no seedlings survived to regenerate the forests.

We now know some of the locations of the Lakeland axe factories, and it is plain that the pioneers swiftly gained an intimate knowledge of the qualities and occurrences of the different local outcrops. Also, we know that the axes were not made in a piecemeal or haphazard way for local use, but were systematically produced to serve an eager market which embraced the Yorkshire Wolds, East Anglia, and Wessex. The most important of the factories so far discovered lay upon the screes of Pike of Stickle, which towers above the head of the Langdale Valley. While hundreds of roughed-out axes

The domed summit of Pike of Stickle is seen directly above the Herdwick sheep. Neolithic stone-axe makers worked on the screes just below the summit.

have been gathered from the Langdale screes, the main concentrations of finished and finely-polished axe-heads have been found on the Cumbrian coast, where the rough-outs must have been carried for the later stages of their manufacture. Some of the products might then have been exported by sea, others by land – some perhaps along the vertiginous routeway known as High Street, which follows the watershed between Haweswater and Hayes Water. The trackway, which in some places is more than 2,000 feet above sea level, was knocked into shape by the Romans, but was probably ancient when they discovered it.

Pike of Stickle is a strangely-formed summit of a volcanic tuff which contains crystals and fragments of quartz, feldspar, iron ore, and epidote. Not a graceful mountain, with its helmet-like upsurging peak, it reaches 2,323 feet, but seems much higher. The best views are gained not from the Pike itself, but from high up on the ridge called The Band, which faces the Pike across Mickleden valley. As I was pausing for breath and a quenching bite of snow on a recent visit to photograph the axe-making site, scale was suddenly injected into the wonderful panorama by the dot-like and slowly-moving figure of a scree-runner, probably a climber risking all to beat closing time at the Old Dungeon Ghyll Hotel. The dwarfing of his figure and my own aching legs served as reminders of the tough determination and

Castlerigg stone circle.

The upright slab in the foreground is Long Meg, and her buxom Daughters form the stone circle beyond.

organization of the Stone-Age craftsmen, who chipped rough-outs in a rock shelter above the windswept, precipitous scree slopes, transported them to the distant coast for grinding, and thence despatched them across the length and breadth of England. Since the discovery of the Pike-of-Stickle site, other axe factories have been recognized in the Scafell–Langdale area. Axes of a different composition were also made in the south of the Lake District, somewhere close to the Whinfell Ridge, using a fine grained mica-rich sedimentary rock known as a 'greywacke'.

While the stone axes were indispensible in the agricultural and economic lives of the New-Stone-Age Lakeland communities, stone also took on a central role in their spiritual existence – in the construction of burial cairns, and particularly in the making of stone-circle temples. One great circle, Castlerigg, built of ice-dumped boulders which were hauled to crown the spur of Cheshunt Hill to the east of Keswick, lies near the core of the district, while other important circles flank Lakeland. These include Grey Croft near Seascale to the west, with ten stones standing up to seven feet tall, weighing up to four tons, and with a diameter of eighty feet; Elva Plain near Cockermouth to the north, with fifteen surviving stones, the tallest less than four feet high, and a diameter of one hundred feet; and, to the east of Lakeland near Little Salkeld village, the vast temple known as Long Meg and her

Daughters – in legend, a coven of witches petrified by the intervention of a saint. Here, the stones are arranged in the form of a flattened circle, with a maximum diameter of some 360 feet, and, of an original membership of around seventy, fifty-nine stones survive, only twenty-seven of them still standing erect. Long Meg herself forms an angular outlier, twelve feet tall and slab-like, rising above her buxom daughters.

Here we have a suggestion that stone *itself* had a place in the ancient religion, for while the members of the circle are rounded, local grey granitic boulders, weighing up to twenty-eight tons, Long Meg is a great slab of red sandstone, hauled from the west over a distance of at least one and a half miles. The choice of an outlier of so contrasting a hue and form hints that the qualities of different stones may have featured in belief. One of the faces of the blood-red pillar is engraved with concentric circles, a spiral and a cup-and-ring mark. Carvings of the latter type are quite commonly found both on natural rock outcrops and on prehistoric monuments; they date from the New Stone and Bronze Ages, but their meaning is quite unknown.

While Long Meg and her Daughters is the grandest Cumbrian circle, the monument lies aside from the tourist trails, and is therefore less celebrated than it might be. Castlerigg is almost as grand, if more compact, with thirty-eight massive members, and a diameter of a hundred feet. It is located in a setting unrivalled by any other English circle. It draws me like a magnet, and its character changes with the subtleties of light and season – haunting at sunset, but perhaps most beautiful in snow and sunlight. The winter rambler who is prepared to brave the snowdrifts and trudge to the deserted circle when the roads are closed and the dazzling, pristine flurries ripple around the stones is promised an unforgettable sight.

Lakeland contains several other stone circles: a few are quite imposing, like lovely Swinside on the margins of the area near Broughton in Furness; several are smaller, with stumpy stones scarcely knee high; and one or two are marked on the maps, but seem more likely to be debris from fallen prehistoric huts or tomb enclosures.

The great religion which the circles commemorate is lost to us, but the effort involved in the construction of the larger circles shows that it was a compelling one, which mobilized the labour forces of extensive areas. For example, it has been estimated that at least 120 men must have strained to erect the larger of the Long Meg boulders. Like the axe trade, the stone circles overlapped the New Stone and Earlier Bronze Ages, and Aubrey Burl has pointed out that the Lakeland circles often seem to be associated with routeways which may have been used in the export of stone axes.* It is even possible that the British fascination with circles began in Cumbria,

**The Stone Circles of the British Isles* (Yale, 1976)

which is well stocked with boulders for circle-making, and that the cult was exported with the axe traders to other British localities.

The religious interest in circle temples declined in the course of the Bronze Age, while the successful removal of the forest and, eventually, the development of more economically-priced iron tools served to diminish the importance of Lakeland stones – though farmers and shepherds still had to co-exist with the rocky landscape of acid, rain-leached soils, barren uplands, and looming crags. Stone remained important in the construction of huts and enclosure walls. The tumbled debris of ancient huts and paddocks (mostly Romano–British or Iron Age in vintage, but some as old as the Bronze Age), can be glimpsed in several remote corners of Cumbria. Some are near the heart of Lakeland, but the most intact and remarkable examples lie on the fells to the east of the district. At Ewe Close near Crosby Ravensworth, Cow Green nearby, and several other sites on the flanks of the western Pennines, one can recognize the stone footings of round and rectangular dwellings, pens, paddocks and field walls.

Field walls of stone rubble are an integral feature of the Lakeland landscape, providing a sense of scale, relieving the bold sweeps of the lower slopes, and defining the boundary between the open, upland ranges and the low-lying, more marshalled farming areas. Not only did they delimit individual properties and the township commons, and exclude the livestock from the hay meadows, former 'infield' granaries, and occasional 'outfield' croplands, but they also provided useful depositories for stones gathered from the boulder litter of intended ploughlands and pastures. At Wasdale Head near Wastwater, 'clearance cairns' of gathered boulders can be seen scattered amongst the walls, which were built much later to enclose the old communal infield.

The Lakeland walls are of many different ages. Some doubtless stand on footings set down in prehistoric periods. Few, if any, will have been built by Romans: the imperialists seem to have been content to keep the natives contained and in check, organizing a few roads across the margins of Lakeland, and building the forbidding stone fortress on Hardknott to control the vital pass on their route from Watercrook, near Kendal to Ravenglass on the Cumbrian coast. Many walls must have been established by the Norse Viking settlers who arrived here via settlements and pirate stations in Ireland, and who will have met with landscapes to remind them a little of those of Norway. We can imagine them putting their well-tried skills to work in clearing the intended meadows and oat and rye fields of boulders, and building pens, enclosures and farmsteads. Some of the latter, with Norse place-names like Braithwaite (wide clearing), Stonethwaite (stony clearing), and Rydal (valley where rye is grown), were eventually destined to grow from farmsteads into hamlets or villages.

The field walls at Wasdale Head near Wastwater serve as repositories for the stones gathered from the fields. The fields are punctuated with stone heaps, or 'clearance cairns', and the walls sometimes broaden to incorporate a mass of stones.

Stone picking and wall building continued through the medieval period, when much of Lakeland was partitioned between the monasteries – ten or so local foundations like the Premonstratesian abbey of Shap, and powerful outsiders like the Cistercian abbey of Fountains – and feudal magnates like the Percys, Huddlestons, and de Lancasters. In the process, great fell-striding walls came to define the vast sheep pastures, and massive stone pounds were built to hold the sheep gathered for shearing or despatch to market.

The Dissolution of the Monasteries provided new opportunities for the class of small, independent farmers, known in the Lake District as 'statesmen', to expand their operations. Before long, however, the pressures of population drove these rugged yeomen to colonize the sterner slopes and lofty backwaters. Some of the decaying head dykes mark the limits of their attempted expansion, though other crumbling walls date only from the nineteenth century, and were built to supersede the boundary stones defining areas of common attached to different townships. Although most are almost impossible to date, there is nothing haphazard about the patterns of Lakeland walls – and the landscape would be much poorer without them. Each farm used, and often still uses, stone walls to define the different types and qualities of its land. At the foot of the unimproved rough pasture of the open fell, a wall marks the boundary of the 'intake', the lower wall-divided pasture attached to a particular farmstead, while the walls of the fields around the farmstead delimit the 'inland', a small complex of walled hay meadows producing winter fodder, which may, from time to time, be ploughed to yield a crop of roots or oats. Other walls define the droves and tracks leading up to the fell, or down to a bridge or to the market.

The statesmen are gone, now, their places mainly taken by bigger landlords and their tenants, but many of their homes, no less rugged than the hill farmers themselves, often survive. A good proportion of the surviving stone-built farmsteads date from the latter part of the seventeenth century, the period when the statesman class was expanding in the aftermath of the Dissolution, and colonizing the new valley footholds. The simple, sturdy dwellings testify to the abundance of stone in the Lakeland pastures and scree slopes, to the hard, intractable nature of these stones, and to the unadorned simplicity of the old way of life. The simplest of these farmsteads have the form of a long-house, a dwelling evolved from medieval predecessors in which only a timber partition separated the family from its livestock. Around the fifteenth century, a stone chimney stack was often introduced at the partition, so that the human and animal quarters would

Raven Crag, with Middlefell Farm at its foot. Above the farmstead are the walls which define the 'intake'.

be separated by a hearth and a stone dividing wall. In due course, the roof might be raised to provide an upper storey of bedchambers, freeing the ground floor for a parlour, a kitchen and dairy accommodation.

Some farmsteads date from the eighteenth and nineteenth centuries, and so incorporate newer tastes, while in other cases the original form of a seventeenth-century statesman's long-house may be difficult to detect amongst the fabric of later additions. Some of the larger seventeenth-century dwellings were built with open, upper-storey spinning galleries, reputedly the places where the housewife and children would sit spinning wool in the bright daylight to supplement the slender family income, but which may mainly have been used for drying fleeces. Others have been altered by the replacement of their open, outdoor stone steps with an 'outshut' addition to enclose a staircase, or enlarged with added kitchens, dairies and elevated or extended barns.

The men who built these farmsteads could not afford imported stone – and certainly not the prestigious limestones and sandstones which could be sawn to display smoothly-squared 'ashlar' faces. Even had they been able to, the stone wagons might never have penetrated the rocky, tumbling Lakeland roads. In consequence, the dwellings and barns were built with cheap practicality, using locally-available materials, and thus they sit snugly in the stone landscapes of overlooking crags and encircling field walls. The rugged, angular slate or slatey slabs of the Skiddaw and Upper Slates and the Borrowdale Volcanics could not be neatly cut or shaped; instead, they were laid in thick, rough courses, with the mortar deeply recessed and invisible, or, as with drystone walling, with massive cornerstones or 'quoins' to bind the four walls together.

The Skiddaw Slates are usually poor roofing materials, but good roofing slates were available from the quarries on Honister Crag east of Buttermere, and in the Borrowdale Volcanics, while various slate quarries exploited the Upper Slates, notably at Kirby-in-Furness. Stone-built Lakeland chimneys may be square or round, and a common feature is the pair of slates forming an apex above the vent, providing some protection against the frequent deluges and snowstorms. Another regional preference is for coating the farmstead walls in white roughcast to improve the weather-proofing, while leaving the adjoining barn in its natural hue.

The buildings of the Lake District thus mirror the resources of their localities in close detail. In the west, where the granite is exposed, pinkish granite boulders often form the walls, with each course of rounded boulders set upon a narrow, stabilizing band of slate. Where these quarries have been at work exporting special stones beyond the district, their products, in the form of massive quoins or lintels of stone – like Shap granite or Elterwater slate – may be seen diverted into the quarry workers' cottages.

A row of industrial cottages dating from the nineteenth century at the slate-quarrying village of Elterwater in the Lake District.

The Lakeland farmstead witnessed little but poverty, combatted by resourcefulness and self-sufficiency, and so it took shape as an offspring of its rugged, rocky environment. The deep shadows in the gaps of its rough-stone courses were mirrored in the crevices and joints of the background cliff-face, the green mosses sealing the roof slates echoed the tints of the surrounding fells, while its stones came from local stock displayed all round in the stream sides and screes. Comforts it might have lacked, but harmony it enjoyed in plenty.

Equally attractive and harmonious components of the man-made landscape are the little, arch-backed, roughstone bridges, which look exceedingly old but probably date mainly from the seventeenth and eighteenth centuries. Ashness Bridge on the lane to Watendlath provides a focus in one of the most delicately composed vistas the Lake District can offer; the bridge in Watendlath itself, and others at Throstlegarth and Middlefell in the Langdale valley, are no less enchanting. Sometimes built on pack-horse routes across the fells, but often just linking a farmstead to its market or providing a safe stream crossing for driven sheep, these narrow arches have proved the downfall of many an alien motorist, but nobody would wish to see them go.

The Lakeland churches also form a quite distinctive group. In contrast to those of the plump, long-settled lowland villages, they seem on the whole to be small, simple, even austere, and relatively youthful. In old Lakeland

– with its scattered population, vast parishes and scarcity of villages, most of them late developers – medieval or older churches were thinly spread. Many small communities lived remote from their parish church. Some managed to scrape together the wherewithal to construct a simple chapel, but the richly-endowed, ornate and often majestic medieval churches so common in lowland England are scarcely known in Lakeland. As W. G. Hoskins explains: 'Although this was a potentially rich countryside, with great sheep pastures, it was owned by other people. The money, the great profits, were drained off and the community was left with the remains. Hence the small poor churches, built with its own money and nobody else's.'

Still, the rocketing towers and ornate embellishments of a palatial Somerset or Northamptonshire church might sit uneasily amid the rugged natural splendour of Lakeland. The existing churches are both unobtrusive and in keeping with the spirit of the place. A good proportion of the region's churches date from the sixteenth and seventeenth centuries, by which times most other English parishes were long since provided with their own places of worship, and church building was at a low ebb. Martindale Old Church, built of roughstone in 1633, provides a typical example of a low and towerless dale chapel; its charm is a product of its austere simplicity. The whitened dale chapel at Mungrisdale dates from the following century and is similarly unadorned. The church overlooking Buttermere village is still more youthful, but no less attractive in its reticent plainness.

Although the ancient rocks of the Lake District have been pierced by a wide variety of mineral veins during many different geological traumas, there are few rich pickings to be made, and the scenery is but little scarred by mining. Such minerals as exist were not exploited till comparatively late; mining and smelting, which tapped the small remaining resources of timber, only really developed in the Elizabethan era, when German prospectors, engineers and miners were encouraged to settle in Keswick – much against the locals' wishes. For a while, the copper smelters built at Brigham were held to be the finest works in Europe, but all that remains of the industry is the rock-cut tunnel which carried water to power the wheels. Ironworking, graphite-mining and slate-quarrying have all helped to supplement the meagre incomes of the district, and some quarrying still continues. The main environmental threats today, however, derive from possible requests for more reservoirs by the thirsty southern conurbations, the further spread of the wretched fell-darkening plantations of alien conifers – and also from ourselves, when we choose to drive rather than walk in this ramblers' paradise.

Chapter 2

Sandstone Scenery

Sandstones are formed from the eroded debris of older rocks, with the rock particles compressed and welded together in a matrix of natural cement. This similarity apart, they are a very mixed bunch. The oldest sandstones began to form when the agents of weathering attacked the original igneous rocks of this planet, long before the dawn of life. The youngest are still in the process of forming, as the weight of accumulating sediments above compresses and compacts the shingles, sands or silts of former beaches and river floodplains. In texture, the sandstones range from the most finely grained siltstones, shales and clays, through the rough gritstones, the knobbly conglomerates or puddingstones with their rounded pebbles the sizes of eggs or heads, to the angular breccias which contain the unrounded debris of old rock screes. The colours, too, are varied. Some sandstones are almost white, pale buff, or yellow, others are iron-stained to rust or cinnamon hues or tinted green by Glauconite minerals, while the London Clay of the lower Thames Valley is blue-grey when freshly dug, but turns brown after its iron content has been exposed to the atmosphere.

On the whole, the sandstones do not produce particularly dramatic scenery on the broad or regional scale. However, locally spectacular scenery can be seen – for example around the gritstone and shale Cliffs of Moher in County Clare, the old Red Sandstone conglomerate cliffs around the knuckle of North East Scotland, or the weird Millstone Grit 'tors' which outcrop at places like Great Alms Cliff and Brimham Rocks in Yorkshire. Because of the many varieties of rocks involved, it is not easy to generalize about sandstone scenery. Few sandstones are as rugged as the tough Millstone Grits, or as unyielding as the 'grey wethers' or sarsen stones displayed in the circles at Stonehenge and Avebury, and so they are likely to form prominent hill or scarp scenery only when they are flanked by still softer and more easily eroded beds. The Greensands of the Weald do provide some of

The gritstone landscape of Brimham Rocks, near Ripon.

the higher ridges in this distinctive 'scarp-and-vale' landscape, and while the local Greensands tend to be soft, they are reinforced by narrow beds of a rather flint-like rock called 'chert'. In general, scenery formed on the coarser sandstones often presents a rather scrubby and parched appearance, which is most clearly and attractively apparent in the retreating sandy heathlands of southern England.

When the parent sandstone rock is broken down by weathering, the finer and less stable mineral constituents may be flushed out of the debris, leaving the coarse and more stable quartz grains to form a free-draining, sandy and rather acidic soil. Sandstone landscapes, however, do not tend to be as dry as those which develop on the limestones; for while rainwater may rapidly seep through the sandy soil, it does not penetrate the bedrock as easily as it would a limestone, and thus surface springs are common features of the sandstone scene. (Though drier than those which are formed of sandstones and often littered with bare rock exposures, limestone uplands do tend to be a fresher green. This results from the varied profusion of calcium-loving plants which thrive on limestone, while the more acidic sandstone uplands tend to be clothed in heather and bracken.) A closer examination of the sandy heathland flora will reveal flowers like the yellow tormentil, blue milk-wort, and the parasitic dodder, while the damper ground may support the marsh gentian, with its sky-blue trumpets, or the brown beak sedge.

The sandstones are also a very mixed bunch when appraised as building stones. Some, like the greenish grey Pennant Sandstones of South Wales and the southern fringes of Gloucestershire, and the Pennine Millstone Grits, are much used locally in buildings of many ages. The best and most evenly-textured were cut as freestones and exploited by Dark-Age and early-medieval stone carvers. Others, like many of the sandstones of the English Midlands, were only deemed adequate for the building of an unostentatious village church or manor house – while others still are virtually useless. These include the soft, flaky shales and the plum-pudding textured conglomerates (although the Saxons did not scorn puddingstone in the construction of North Elham cathedral in Norfolk, and a reddish iron-conglomerate is displayed in several lesser Saxon churches).

Geologists regard any sedimentary deposit as a rock, whether it be firmly consolidated, or loose and unbonded like the sand in a dune, or soft and sticky like clay. In this context, clay can be regarded as the most successful building rock of all, since clay bricks have ousted most other stones from the twentieth-century building sites. Sandwiched between beds of tougher chalk or sandstone, beds of clay are responsible for forming the damp vales which alternate with scarp and dip-slope scenery in south-eastern England. Many clays, like the stiff blue Gault of the Weald, originated as very fine rock particles which drifted far out to sea, settling on seabeds to form marine muds. They are easily eroded to form low, rather featureless vales. The minute and densely-packed particles resist the passage of water, and thus

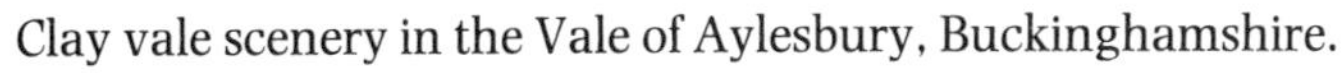
Clay vale scenery in the Vale of Aylesbury, Buckinghamshire.

The scenery of brick-making near Stewartby in Bedfordshire.

the claylands tend to be waterlogged and slippery in winter, while the stiff, clinging soils are slow to warm in the spring. As a happy consequence, some clay vales still succeed in repelling the ploughman, and support green pastures and moist flower-rich meadows beside slow, shaded streams: clay vale scenery is not spectacular, but it often contains peaceful, verdant details.

Those who have a liking for surrealistic landscapes may appreciate the scenery of Bedfordshire brickmaking. Not many people do. The county is carpeted by a variety of different clays: there is some Gault; thick swathes of glacial boulder clay; Clay-with-Flints – a sticky red clay studded with flints deposited during the erosion of the chalk mass which contained them; and also beds of grey-blue Oxford Clay. A clay-brick industry has existed in Bedfordshire continuously since the fifteenth century, but it was the development of the Fletton industry which transformed the landscape of the central vale of the county. At Fletton, near Peterborough, it was found in 1881 that beneath the surface of the Oxford Clay lies a greenish, shale-like deposit, the 'Knotts'. While the visual attractions of Fletton bricks, in comparison with the older russet products which are displayed in many seventeenth- and eighteenth-century buildings, may be debatable, in other respects the Knotts provided an ideal brickmaking material: this clay can be crushed, pressed and fired directly, without a curing stage; it contains combustible material which allows enormous economies in fuel; and it includes just sufficient lime to prevent the bricks from cracking during firing, while containing few impurities.

The common at Brill in Buckinghamshire is pockmarked by clay pits, a legacy of the pottery, tile and brick-making industries which existed here from the early Middle Ages until quite recent times.

In 1926, work began on the construction of the 'Garden Village' of Stewartby, purpose-built in new Fletton 'Rustics' to house the brick workers and named, with as much modesty as was thought fitting, after the company chairman. The surrounding forest chimney landscape and open clay workings are fearfully obvious in the otherwise featureless landscape of clay vales, but at least they contain considerably more interest than the modern prairie fields which increasingly deface the countryside around Bedford.

One of the most attractive and individual of the sandstone landscapes is associated with the Millstone Grit of the Pennines. The mountain and upland spine of England is composed of two distinctive and very old rocks: the Mountain Limestone and the Millstone Grit. Firstly, around 400 million years ago, the great thicknesses of limestone were deposited upon seabeds, and then enormous accumulations of grit were laid above the limestones, in shallow waters which were swept by powerful currents. Then the seafloor deposits were uplifted and erosion began its remorseless work. In places, great slabs of Millstone Grit remain as isolated caps to form the summits of mountains like Ingleborough, Pen-y-Ghent and Whernside, which rise above the surrounding limestone. Elsewhere, millions of years of erosion, faulting, and tilting have masked the original relationship between the limestone and grit, and the Millstone Grit exists in detached sweeps and ribbons of country. It provides the scenery in the eastern section of the Yorkshire Dales, parts of the Derbyshire Peak District, a block of country around Lancaster, and narrow approach zones to the Pennine limestone in the valleys of the Tyne and Tees. Perhaps the most dramatic of the grit outcrops is the rugged scarp of the Roaches on the Staffordshire margins of the Peak District.

The gritstone scenery forms striking contrasts with the adjacent Mountain Limestone landscapes, for these are scarred and dappled with pale rock outcrops, support few streams, but carry a bright carpet of pasture. If the gritstone scenery is generally less distinctive, it has its own darker, brooding and almost soulful quality when it forms the moors which loom above the wall-striped patchwork of valleyside pastures and hay meadows. For some reason, the popular perception of the Yorkshire Dales is founded on the limestone fastnesses of areas like Malham, Grassington and Wensleydale, but the grit is an equal partner in the Pennine landscape, and Nidderdale is as attractive as any of the celebrated limestone dales. Like most other coarse sandstones, the grit breaks down to produce a soil which is sandy, acidic and rather poor. Valuable minerals and humus tend to be washed down through the quartz grains which remain. In consequence, the swelling gritstone slopes are seldom ploughed, though a reasonable pasture results when lime from the adjacent limestone country is spread upon the fields.

If the gritstone scenery is not so bright a green as that on the limestone, at least it can claim to be more lush. The rivers and streams of the limestone areas which emerge from their caverns to pursue slender, perilous courses across the thirsty rock are revitalized when they cross into the grit. Tributary streams swell and multiply, while alder, oak and elm now shade the banks, following the course of the rivers in bobbling green ribbons which weave and wander along the dale.

With its smoothly-swelling slopes and undulating plateaus, sandy soils and ling-carpeted moors, gritstone country is in some ways reminiscent of a granite landscape. The resemblance does not end here, for many a swell is crowned by the tottering silhouettes of tor-like features – and Brimham Rocks near Ripon are as grimly outlandish as any Dartmoor tor. Although their origins are completely different, the Millstone Grit is coarse-grained, like granite, and it is also deeply fissured by joints and prone to fracturing and weathering along horizontal bedding planes. While the gritstone tors may have formed in a similar manner to those of the granite landscapes, grit is not quite as hard as granite. On close inspection, these knobbly, blocky outcrops can often be seen to be grooved by series of smooth, parallel furrows. This gouging is the work of windblown sandgrains which have gradually exploited slight differences in the resistance of different layers of strata within the rock. Close study may also reveal a little more, for one can often see the evidence of 'current bedding'. The angular grains which formed the grits were deposited in shallow, quite turbulent water and, as the land sur-

One of the grit tors at Brimham Rocks.

face eroded rapidly, great submarine fans of debris must have accumulated. Thus, instead of being deposited in horizontal beds, the grit may have been dumped to form sloping layers, like debris tipped from the top of a slag-heap, while sea currents swept and swirled the sands on the seabed.

As in Dartmoor and the Lake District, the underlying rock firmly asserts its presence in the landscape. Although not widely used in noble buildings lying outside the Millstone Grit territory, it has been used locally – in the building of churches and great houses, in the nineteenth-century terraces and public buildings which fossilize the former glories of the fashionable spa of Harrogate, and in the more humble and soot-blackened terraces of a clutch of Pennine mill towns. With its many swift streams, the gritstone country offered attractive sites for the water-powered mills which flourished in the days before the adoption of steam power drew industry towards the northern coalfields. The grit also provided the materials for the dozens of slender-arched pack-horse bridges used for market trading and for dispersing the products of the earlier mills. Although, like the Lake District bridges, they often seem to be positively ancient, most of these bridges belong to the seventeenth, eighteenth and early nineteenth centuries, and some replaced older, flood-prone wooden bridges. The use of gritstone in the building of the much-admired rugged farmsteads and cottages of the area was also a post-medieval practice, and the eighteenth-century traveller in the Dales would have seen many cruck-framed cottages, which preserved an older vernacular building tradition, but which are now rarely seen. Finally, there

A rugged gritstone church at Hubberholme in Wharfedale.

are the networks of field walls of many ages, containing darker and more blocky stones than the angular slabs which provide the walls in the neighbouring limestone areas.

Although one may cringe at the modern 'improvements' which afflict many old farmsteads in the Dales – with the vogue for large 'Georgian' bow windows not the least offensive – at least the authorities have required new private dwellings to employ facings in the 'traditional' stone – Millstone Grit or softer sandstones from the coal-mining areas. Thus the visitor has many opportunities to compare the honey and cream-toffee tints of the freshly quarried grits and sandstones with the dull olive greys of the weathered stone. The enthusiasm for cleaning buildings in the soot-encrusted mill towns has created some striking juxtapositions of black and buff. The value of these cosmetic undertakings is a matter of taste. To the outsider, it may seem as though the townsfolk are seeking to purge the environment of the stains and scars of their noble – if squalid and arduous – industrial past.

While other rocks – like the slate of the fells of Ravenstonedale, the shales and grit of the Yoredale beds which underlie the Millstone Grit, and the sandstones, shales and coal seams of the Coal Measures – make occasional appearances in the Pennine Dales, in essence the landscape is a combination of the different personalities of the Mountain Limestone and Millstone Grit. Sandstone of a remarkably ancient type is a partner in another distinguished and dramatic landscape, which is also geologically schizophrenic. The area concerned lies along the extreme north-western fringes of the British mainland, and is a part of Scotland which has been geologically separated from the remainder of the country by the great fault known as the 'Moine Thrust Plane'. The rocks which form the scenery are the Lewisian gneiss and the Torridonian sandstone. Both belong to the most ancient geological period, the Pre-Cambrian, and they are among the oldest rocks in the world.

The hard grey gneisses had experienced a long and complicated existence before life on earth even began. The original rocks, perhaps volcanic ashes and lavas, were thrown up into mountain chains and metamorphosed into gneisses between 2,000 and 3,000 million years ago. Then, swarms of dykes consisting of a basaltic magma were injected into the already ancient rocks, and a second great era of mountain building contorted them in the period up to about 1,500 million years ago. In due course, the forces of erosion began to level the sinking margins of the ancient continent of which the

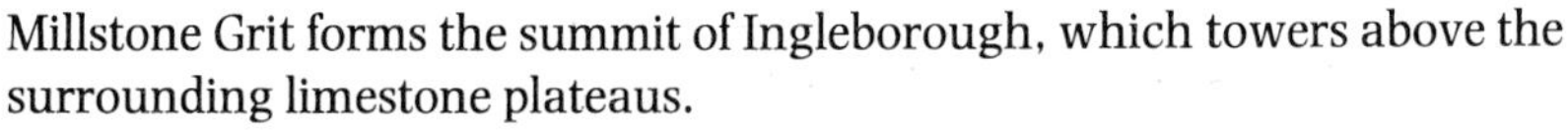
Millstone Grit forms the summit of Ingleborough, which towers above the surrounding limestone plateaus.

The Torridonian sandstone mountain Stac Pollaidh standing on an ancient plain of Lewisian gneiss.

Lewisian gneisses formed a part. The old continental surface became blanketed under layer after layer of river sediments – sandstones, mudstones and conglomerates – which accumulated to thicknesses of up to 20,000 feet as the old land mass sagged and sank.

Subsequently, erosion has progressively stripped away the Torridonian sandstone cover and so the old land surface of the Lewisian gneiss is reappearing. The landscape which results is a strangely beautiful one that resembles some artists' conceptions of a Martian or Venusian scene. Massive isolated blocks of plum-red Torridonian sandstone form humped or conical mountains which tower above the knobbly undulations of the gneiss plain, and the ghostly, fairyland effect is often heightened by the West Highland mists and hazes which so often shroud the mountain bases. Formerly, the sandstone was covered by a protective sheet of tough quartzite. This was gradually eroded into islands, providing protective cappings on the summits which remained as peaks, while the surrounding sandstone was gouged away. The pinnacle of Stac Pollaidh which towers above Loch Lurgain has now lost its quartzite shield, though the taller peak of Suilven to the north is still protected. Those who ramble across the Lewisian gneiss tread on an exhumed landscape, a surface which existed millions of years before the dawn of life.

Chapter 3

Landscapes of Limestone

The limestone family embraces a range of quite distinctive rocks, but virtually every one provides the basis for delightful scenery or attractive buildings. All the limestones have calcium carbonate as their main constituent, and this lime generally derives directly from the tiny skeletons or shell debris of ancient sea creatures, or has been precipitated from lime-rich water to produce less common rocks, such as calcite. As with the sand-stones, the visual, structural and age range is great, and the limestones vary considerably in their value as building materials. Most are light in colour. If we include chalk within the limestone family, the spectrum runs from snow white, through the creams, buffs and iron-stained light ambers of the Jurassic stones of the Cotswolds and Northamptonshire Uplands, to the silver-grey of the Mountain Limestone. The spectrum does not end here, for in parts of the west of Ireland such as the Darty Mountains, one sees steep-edged plateaus of blue limestone which seem from a distance to be formed of basalt sheets. Finally, there is the dark, shell-spangled limestone of the famous Purbeck marble.

Limestone tends to produce bold scenery. In the cases of some of the older members of the group, like the Mountain Limestone, this is largely a reflection of the toughness of the rock. Chalk, the youngest and purest of the lime-based rocks, is among the softest of stones, but even so it forms some of the highest scarps in the undulating landscapes of southern and eastern England. This is due to a number of causes: the fact that the Alpine era of mountain building, whose storm waves rippled and folded the young rocks of south-eastern England was relatively recent – ten to twenty million years ago; the softness of the rocks which border the chalk; and the ability of chalk to absorb many of the rivers which would otherwise have gouged and lowered the downlands.

Limestone landscapes are characteristically dry. The rocks tend to be fissured by networks of vertical cracks or joints. These joints have enabled

The limestone beds at Lulworth Cove in Dorset display the effects of folding during the Alpine era of mountain building.

quarrymen to extract large blocks of stone with ease, but they have also captured the runnels of rainwater which would otherwise have merged in networks of streams and rivers. Falling rain becomes a very mild form of acid as it takes carbon dioxide and minute quantities of other chemicals from the atmosphere, and the acidity of rainwater can increase as it accepts other impurities when trickling downwards through the soil. Although limestone rock is often hard, it can gradually be dissolved by exposure to acidic water. Thus, water which percolates through the vertical joints and along the bedding planes of the rock will eventually carve great underground tunnels, caverns and swallow holes. Few surface streams can travel far before cascading down a joint to enter this underground labyrinth of waterways. Most of the rock may dissolve, and such soil as forms on a limestone surface tends to be flushed away by the disappearing runnels and streams, so limestone surfaces tend to be thinly soiled or bare. Meanwhile, the removal of the soil blanket and the chemical attack on the joints and crevices of the bared limestone surface may produce the crazed pattern of slab-like 'clints' and crevices or 'grikes' of the typical limestone pavement.

This extreme type of limestone scenery, known as 'Karst', is associated with the older, harder and relatively pure rocks of the Mountain Limestone. Although dryness is a characteristic of other limestone landscapes, the Jurassic uplands of the English Midlands do not display the craggy scars or bare pavements of the Pennine mountain country. The Cotswold, Dorset and Northamptonshire limestones are less tough than their Pennine cousins, and erode more rapidly to provide a surface blanket of soil and subsoil which can then be bound and stabilized by pasture or woodland. Rather than allowing the agents of chemical attack to concentrate on assaults along joints and crevices, chalk will readily absorb rainwater, which passes freely through the porous rock until it reaches either the underground water table,

These honeycomb patterns have been produced by the erosion of limestone in the cliffs at Elgol, Skye, by the action of waves and salt spray.

or an impermeable sheet of strata beneath the chalk. While the limestones are associated with landscapes which are generally, though not entirely, lacking in surface streams, the different rock types produce scenery as attractive and diverse as that which can be seen in the upper reaches of the Yorkshire Dales, the rolling uplands of the Cotswolds, the cliff-girdled plateaus of County Sligo, the sweeping downlands of southern England, or the barren fastnesses of the Burren in County Clare.

Most limestones are rich in fossils, and because the different types of animals from which they are formed occupied different ecological niches within the ancient marine environments, marked local differences are found in the composition of rocks which are of roughly the same age. The various sea creatures whose shells and skeletons contributed to the formation of rocks tended to thrive in shallow and often rather still waters, and some limestones represent the fossiliferous remains of offshore reefs. Reef limestones occur in beds of many different ages, and amongst the oldest are those of Wenlock Edge in Shropshire, a cliff-like escarpment of grey rocks which are up to 500 million years in age. Reef deposits also occur in the limestone outcrops of south-western Devon; they resemble the Mountain Limestone of the Pennines, but are older, and sometimes of a pinkish hue. More than 400 million years in age, this Devonian limestone provides the basis for some splendid sea-cliff scenery, and for many of the older buildings of Plymouth and Torquay. Coral-based limestones can be seen in the Mendips, and among the oolites and ragstones of the Cotswolds. The shells of many ancient sea creatures – ammonites, oysters and other molluscs – occur in a variety of British limestones, but perhaps the most distinctive are the crinoidal limestones. These can be found in County Sligo, northern Lancashire, and in the limestone beds of Derbyshire and South Wales, and may consist entirely of the closely-packed calcite stalks of sea lilies. The tiny ooid and coccolith skeletons of oolitic limestones and chalk respectively naturally give rise to rocks of a much finer and more even texture.

Less widespread are the limestones which result from the solution of existing calcareous rocks and the precipitation of the lime to form a crystalline limestone. Calcite is such a rock. Most readers will have seen it in the forms of stalactites and stalagmites. The downward-pointing stalactites form very slowly as lime-rich water droplets drip from a cavern roof, depositing minute specks of calcite which gradually grow in icicle-like formations. Meanwhile, stalagmites grow upwards as the drips deposit calcite on the cavern floor. Eventually, a stalagmite and stalactite pair will grow together and merge to form a dripstone pillar. Less natural stalactites form quite rapidly from lime-charged drips from mortar or concrete. Dolomitic limestone resembles calcite, and the tough, yellowish Magnesian Limestone belt is the most extensive outcrop, as well as being a source of excellent building materials.

A dripstone pillar in the spectacular Stump Cross caverns near Pateley Bridge, formed by the union of a stalactite and a stalagmite.

Tufa is a rather soft and friable limestone, rich in calcite, which is precipitated through the evaporation of water around springs or in caves. It will be familiar to all visitors to petrifying springs such as the well at Matlock in Derbyshire, where the tufa swiftly forms a stony coating around the offerings of bric-a-brac.

The most widely known and visited of the British Mountain-Limestone landscapes is surely that of the upper Pennine Dales, but the Somerset Mendips and Dovedale on the Derbyshire and Staffordshire borders are other areas whose limestone scenery is justly renowned. There can be few more impressive introductions to the dramatic qualities of Mountain Limestone scenery than a visit to the surroundings of Malham village in Yorkshire. Just to the north of the village is the great cliff-faced amphitheatre of Malham Cove, a bare, vertical crag almost 250 feet high. Faulting along the line of the mid-Craven Fault produced this great cliff, and originally Malham Beck cascaded down the crag face. Subsequently, however, the stream burrowed a new course beneath the limestone and now it emerges from a cavern at the foot of the Cove. Within walking distance of the Cove, on the other side of Malham village, the great gorge of Gordale presents scenery which is even more dramatic in a heavier, almost oppressive way. The lower section

Characteristic limestone scenery at Malham Cove in Yorkshire.

of the gorge is flanked by steep and overbearing cliffs and seems to have resulted from the collapse and removal of the limestone beds which formerly roofed a great cavern. Near the head of the gorge is a waterfall which dwindles to a trickle in summer as the fissured, thirsty limestone upstream takes its toll of the flow. The scenery around Malham is sure to fascinate and impress the visitor, but it is only fair to add that the little village, and its environs, seems to receive more summer visitors than the area can really absorb – I now prefer to enjoy the landscape in the winter.

The Pennine limestone scenery has many, many more delights to offer. Kilnsey Crag in Wharfedale is an imposing limestone scar which towers above the road running along its foot. The effect is heightened by the undercutting of the base of the cliff and the bulging overhangs. Limestone produces impressive surface landforms, but it also has its own distinctive underground scenery. The caverns and swallow holes of the Dales attract scores of hardy potholers, but less specialized and resolute visitors cannot fail to be excited by the stalactite and stalagmite formations in Stump Cross Cavern, which burrows into an island of limestone set within the scarred landscapes of former lead-mining between Grassington and Pateley Bridge. Then there are the splendours of the area around Ingleton at the foot of the grit-capped peak of Ingleborough – with flashing white cascades like the Holly Bush Spout, Thornton Spout and Beezley Falls, the cave formations of the White

The waterfall in Gordale flowing between deposits of soft, water-deposited tufa.

Kilnsey Crag in Wharfedale.

Scar Caves, and Gaping Ghyll, Britain's most spacious pothole, which plummets to a depth of 365 feet.

Though born and raised in the Yorkshire Dales, I will give a more detailed account of an area of limestone scenery which lies within the British Isles, is much less visited or renowned than the landscapes of the Pennines or the Cheddar Gorge, but which is, in its way, even more dramatic and exciting. The Burren in County Clare in the west of Ireland is a barren, almost desolate expanse of limestone, with karstic scenery so extreme that it almost seems to caricature other limestone landscapes.

The rock of the Burren resembles the Mountain Limestone of England, and it occurs more widely in near-horizontal beds, which are clearly displayed in the ruggedly-stepped hillsides. Plains of flat limestone pavement separate the upland masses, and the surface of the landscape is pockmarked by great hollows, or 'turloughs', which capture the floodwaters from the frequent rainstorms that roll in from the Atlantic. Some turloughs hold permanent lakes, but most fill and empty quite rapidly as the floods seep away through the network of grykes and fissures. After the waters have receded, the turloughs remain as sunken islands of green pasture set in a pale grey of clints and grykes.

Man and nature have combined in the creation of the landscape of barren pavements. Ice sheets and glaciers removed much of the ancient soil cover, depositing massive boulders, or 'erratics', on the denuded surface when they diminished and receded. Even so, without human intervention, the Burren scenery would be much less bare than it is today. As the climate improved during the Ages of Stone and Bronze, pine, hazel and yew forests advanced across the ice-scarred landscape, seeds germinated in fissure-filling pockets of soil, and the Burren was blanketed in woodland. In many of the lowlands, these trees were yielding to a mixed deciduous woodland of oak, ash and elm when pioneer farmers began their assault on the wildwood, clearing land for grazing and ploughing and thus exposing the limestone to soil erosion. Were the grazers to withdraw from the Burren, then hardy trees like the hazel would quite rapidly recolonize the landscape, providing shelter and shade, increasing the humus content of the thin soils, and preparing the stage for a natural re-afforestation. Virtually all the landscapes of the British Isles are in large measure man-made; there are few areas that are more wild and deserted in aspect than the Burren, but even this distinctive wilderness is a product of human intervention.

Within this 370 square miles of apparent desert flourish a number of varieties of flowering plants which are rare or unique and exquisitely beautiful. Some plants which are common on the British mainland did not succeed in colonizing Ireland, because their northward and westward post-glacial advance from the Continent was checked by the opening and widening of the Irish Sea as ice sheets melted and water returned to the oceans. A part of the flora of the Burren, including some Mediterranean species, survives to represent the pre-glacial flora of the area, while the ancestors of some other plants may perhaps have germinated from the seeds of northern flora carried southwards by advancing ice.

A bee orchid, one of the many beautiful orchids which can be seen in the Burren.

The result is a weird botanical wonderland; the warm air from the nearby Atlantic warms the area in winter, when the colder, heavier air sinks to fill the hollows and lowlands. Thus, one encounters warmth-loving plants enduring on the uplands, even though the lowlands are too cold to support them. For centuries, the stockmen have driven their cattle to the higher winter and spring grazings, bringing them down to the lowlands from May to October. Anyone who has walked over the glaring pavements of the Burren on a sunny day will be only too well aware of the heat and brilliance of the setting. The limestone outcrops absorb the warmth of the sun, while the pale, bare pavements and the adjacent sea reflect the light. The warmth and brilliance of the Burren environment provide ideal conditions for many low-flowering plants, which variously bloom throughout the year, although the most colourful displays are seen in late spring and early summer. Many of the plants are lime-lovers, and are tolerant of dry conditions, and hosts of them flourish in tiny, crevice-bound pockets of soil, irrigated by trickles from the frequent showers, or nourished by streamlets which dance in and out of the labyrinth of fissures.

The profusion of plant varieties is quite amazing, and a rapid species count by my wife revealed twenty-six different plants growing within an area of just four square feet. Even the rare species are too numerous to list, but pride of place must go to the orchids, which include the multicoloured fly and bee orchids, the abundant early purple orchid, the fragrant orchid, which is a paler shade of pinkish mauve, and the creamy-white, star-like flowers of the lesser butterfly orchid. Other flowers bloom in a profusion of different colours so that the landscape, which seems from a distance to be composed of almost lifeless grey terraces and pavements, is transformed as one approaches into a brilliant tapestry. There are spring gentians of the

Poulnabrone tomb in the Burren, built of massive limestone slabs.

purest blue, mountain avens and hoary rockrose in shades of white and yellow, the violet tinted buttterwort and white spring sandwort – each plant adding a little splash of brilliance to the galaxy of colour.

If the Burren is a paradise for the botanist, it is no less attractive to the archaeologist, and I can think of no better place for the amateur prehistorian to ramble in search of ancient monuments. The Burren provided the prehistoric tomb or fort maker with an accessible and inexhaustible supply of building slabs which could be prised with ease from the ubiquitous pavements. There are a few portal dolmens (see Chapter 6) larger than the finest Burren example at Poulnabrone, but none displays such an outlandish and haunting silhouette: its great capstone, twelve feet in length, overlaps the clustered supports like the top of a gigantic dining table. There must be at least 150 of the box-like wedge tombs, which may date from about 1500 BC, but the profusion of circular stone forts, dating mainly from the Dark Age, is quite astounding. From some forts, one may be able to see half a dozen others, all within easy walking distance. A few – like Cahercommaun or Cathair Chonaill – are strong and imposing structures, but most of the stone forts are so flimsy that an assailant could have kicked down the chest-high walls. Rather than being castles or fortresses, these are surely the remains of walled farmsteads with the mere trappings of defence.

The stone forts testify to a formerly well-populated Burren, and throughout the area the ruins of stone buildings of many different ages exceed the inhabitable constructions. Although the Burren is still grazed by cattle and by free-ranging herds of goats, the current human population can be only a fraction of that which the area supported 3,000, 1,000 or 100 years ago. Even so, it can never have been a land of plenty, and its name, *boireann* in Irish, means 'rock', or a stony land. The history of Ireland is in part a catalogue of disasters, with local wars, invasions, massacres, evictions and famine each contributing to the ubiquitous heritage of ruins. In the case of the Burren, many of the desertions seem to have occurred in recent centuries, and reflect problems of an environmental rather than a political nature. While most of the surviving Burren farmsteads and villages lie on the low ground, the pattern of ruins shows that the stone forts and many of the medieval churches, lost villages, hamlets and farmsteads were in upland situations. It therefore seems probable that in the course of the historical period an intensification of the process of woodland clearance produced a lowering of the water table, and an acceleration of soil erosion – with disastrous consequences for grazing and tillage.

The ruined stone buildings of the historical period come in many forms. There are the relics of important ecclesiastical centres, such as the cathedral complex of Kilmacdaugh on the eastern flanks of the Burren. Here there is a fine round tower, now leaning precariously but still intact, around which are grouped the remains of an eleventh- or twelfth-century cathedral, a small twelfth-century church, another of the thirteenth century, and yet another which may belong to the fifteenth century, as well as what seems to be a much altered abbot's house. In the north of the Burren are the twelfth-century ruins of Corcomroe Abbey, while on the southern margins lie the remains of the medieval cathedral of Kilfenora along with three high crosses of the twelfth century, and the religious centre of Dysert O'Dea with a remarkably fine high cross. The ruins of lesser medieval churches abound. There are also the remains of deserted medieval villages, like Lias an Ru, and others more recently lost, like Baile na Creige, while the barren uplands are studded with the ruins of abandoned hamlets and farmsteads. There is also a legacy of ruined castles to rival that of other parts of Ireland – the tower house and fortified manor of Lemeneagh and the round tower of Newtown Castle being amongst the finest examples.

Hardly less evident than the bare limestone pavements and scars of the Burren are the lattices of field walls which traverse them, partitioning even the most barren and desolate of the slopes and summits. In the nineteenth century, a rationalized gridwork of larger enclosures with straight walls was

The scattered bones of a cave bear in the Aillwee caves.

superimposed upon the landscape. However, the days when tillage could be practised upon the now-denuded lands were gone, and so there was little incentive to remove the older network of smaller, irregular field walls. Rough and ready walls could be built with ease from angular stone slabs lifted from the pavements, but neat walls of evenly-sized stones, shaped and carefully laid like those of the Cotswolds, are seldom to be seen. The tough and irregular slabs of Burren limestone do not lend themselves to such an orderly treatment. Instead, the field walls of the Burren seem to give free expression to the abilities, skills, or lack of them, and foibles of the various builders. In some, the stones are set in fairly regular horizontal courses, in others they are vertical, while others still adopt a herringbone pattern.

In its underground landscapes, the Burren can compete with the Mountain Limestone areas of the Pennines and Mendips. More than forty miles of cavern network have been discovered and explored in the region, and one of the less accessible caverns contains a stalactite which is more than twenty feet in length, and is believed to be the longest example in Europe. Only one set of caverns is open to the general public, but the Aillwee Cave, with its ample and discreet parking area and unobtrusive restaurant and souvenir stalls, is a model of presentation which the proprietors of some southern English caverns might do well to imitate. The entrance to the cave was only discovered in 1944, and it was not until 1977 that the removal of a choke of boulders revealed the existence of a new cave with vast caverns. The exploration of the further reaches of the network is still continuing.

A fine expanse of limestone pavement above Malham Cove.

In comparison with the Stump Cross Caverns, for example, the stalactite formations of Aillwee, which are stained black by manganese impurities or red by iron, are not particularly impressive; but the impact of the caverns derives from the length and spaciousness of the tunnels and chambers. Most of the gouging and tunnelling was accomplished by meltwater during the dying stages of the Ice Ages. Before melting, the ice dumped erratics of Connemara granite in the cavern mouth. Subsequently, the river which occupied the network cut a lower course, and so the caverns are relatively dry; the small cascades which can be seen derive from rainwater that has percolated down through 500 feet or more of rock. Several thousand years ago, bears hibernated in an area about 200 feet inwards from the cave mouth – their paw and claw marks can still be seen in the cave floor muds, along with the skull and bones of one of the beasts, which are displayed *in situ*.

The Burren is the place which, more than any other here described, I would encourage the reader to visit. Sadly, many potential visitors are deterred by the current troubles in the north, and yet this most lovely and welcoming of islands has so much to offer, and the tourist in the south has nothing to fear. I know of no landscapes more enthralling and no people more agreeable than those of the still unspoilt countrysides of western Ireland.

Before we move on to some of the other limestone landscapes, one conservational point should be made. Limestone provides a basis for many an attractive garden rockery, but some of the rockery stone which is offered for sale originates from areas of limestone pavement. These pavements take centuries to form, and the tiny crevice-filling pockets of soil which support their often rare and colourful plants are also very slow to accumulate. If you find such stone on sale, therefore, and if you care about the future of our stone landscapes, please do not buy it. In 1983, a Nature Conservancy Council report estimated that we have already lost forty-five per cent of our limestone pavement.

The natural dramas of Mountain Limestone scenery are not apparent in the rolling countrysides of the Cotswolds and the other landscapes of the Jurassic limestone belt. Younger, less resistant and less pure than the limestones of the Carboniferous period, those of the Jurassic tend to erode more rapidly to produce a blanket of litter which protects the rocks beneath, while containing material which will persist as soil rather than dissolve like the purer lime. Pavements are not a feature of these landscapes, and bare stone scars, like those at Birdlip Hill in Gloucestershire, seldom feature in the scene. While the karstic Mountain Limestone landscapes are wild and barren, those of the Cotswolds are golden, mellow and mature – just like good Gloucester cheese.

Stunning land forms are seldom seen in the Cotswolds. The limestone beds dip gently towards the clay vales to the south-east, while presenting an often quite steep face towards the north-west. Those who have been part of a tourist caravan grinding southbound out of Broadway in Worcestershire will appreciate the steepness of the scarp. It is largely in the works of man that the stone of the Cotswolds area makes its most obvious and enjoyable contributions to the landscape. Like the majority of dry limestone soils, those of the Cotswolds support a varied and springy pasture, and the area was one of the leading producers of the sheep whose 'golden' fleeces contributed so greatly to the wealth of medieval England. With excellent building stones widely available and cash flowing in from the lucrative wool trade, it is not surprising that the area contains many fine medieval churches. Fairford has one of the best examples, and is rare in having preserved its painted glass. It was provided by John Tame, a wool merchant, at the end of the fifteenth century – and so, like a high proportion of the most splendid of the Cotswold churches, it is built in the Perpendicular style. Chipping Campden was a centre for the medieval cloth trade which finds its most impressive epitaph in the little town's late Perpendicular church, with its tower based on that of Gloucester cathedral.

The oolites and ragstones have also been used in the building of houses great and small. With excellent, amenable materials so easily available, the area contains some of the oldest surviving examples of stone-built vernacu-

The vernacular architecture of the Cotswolds as seen at Lower Slaughter.

lar architecture. Some cottages date back to the sixteenth century and represent the earliest stone phases of the 'Great Rebuilding'. While most of the great houses were designed and built according to the current fashion and are therefore easily dated, the lesser dwellings use a vernacular style which was remarkably persistent. Steep-pitched roofs, lots of gables and gabled dormers, and stone-tiled roofs are the most obvious themes; because the village masons were conservative, reluctant to change a successful and attractive style, the cottages are often hard to date.

Tourists now converge on the Cotswolds, and the popular enthusiasm for the stone-wrought details of this charming landscape is not surprising. Unfortunately (or fortunately, depending on whether one runs a café or knick-knack shop), certain villages, like Broadway, Castle Combe and Bourton-on-the-Water, have captured the limelight and are sometimes unable to cope with the swarms of visitors. The clamour and traffic obscure the very genuine charms of these places. It is strange that so few people realize that all around these congested showpieces there are scores of other stone-built villages which are lovely, interesting and uncluttered.

It is also strange how tastes in landscapes seem to change. The Cotswolds may yield only to the Lake District or Cornwall in the popular conception of 'ideal' scenery. In the 1820s, William Cobbett wrote of the area of walled fields outside Cirencester: '... anything so ugly I have never seen before'. About the same time, Sidney Smith described the uplands of Gloucestershire as '... one of the most desolated counties under Heaven, divided by stone walls and abandoned to the screaming kites and larcenous crows: after travelling . . . over this region of stone and sorrow, life begins to be a burden and you wish to perish.'[2] Since then, we have learned to appreciate rugged or unregimented scenery – although we seem set to destroy it with conifers, synthetic building blocks, telephone wires and factory fields.

Chalk is the most instantly recognizable of the British rocks. The white cliffs of Dover occupy a special niche in the national consciousness, signifying 'home' to many a returning seaman, soldier and expatriate, and also symbolizing the defiance of a nation which has resisted several attempted foreign invasions. The brilliant white hill figures which are carried like brands upon many a downland scarp – such as the white horse of Uffington and the Cerne Abbas giant – are familiar local landmarks. Chalk may form rather featureless and uninspiring boulder-clay smeared planes, like those of part of Cambridgeshire, but it also provides the basis for rolling, open upland scenery. Although the English downlands never rise above the 1,000 feet contour, the North and South Downs, Chilterns and the Downs of Berkshire, Marlborough, Hampshire and North Dorset once provided welcome islands of fresh and bracing countryside, with broad vistas across the intervening vales.

The white chalk cliffs above the harbour at Dover.

The Westbury white horse, one of several such horses which were created by clearing the turf and soil from chalk scarp faces in Wessex during the eighteenth century.

Sadly, the downlands, which provide such welcome refuges from the clamour and ugliness of the industrial south-east or the monotony of suburbia, are retreating. Several chalk uplands, particularly those of Dorset, preserve the rectangular patchwork patterns of prehistoric Celtic fields or the terrace-like strip lynchets of the medieval period. These successful, or desperate, attempts to plough the thin, dry chalk soils were, for one reason or another, abandoned, and the view that the downlands were best farmed as extensive sheep pastures more generally prevailed. Broad tracks of downland, which had existed as permanent pasture for hundreds or thousands of years, surrendered to the plough during the Second World War, when enemy submarines threatened to sever our links with the food exporting countries overseas. Today, for reasons which are less pressing or convincing, governments have chosen to subsidize the ploughing of still wider areas of old pasture. There are few vistas more sterile and dispiriting than the arable wilderness of lost Wessex downland. By 1983, some eighty per cent of sheep-grazed calcareous pasture had been lost, and with it a wonderful heritage of archaeological and botanical treasures.

About 100 million years ago, most of what is now Britain lay beneath a vast ocean, and great thicknesses of chalk were deposited. Subsequently, the land re-emerged and volcanic eruptions took place in the north and

The face of the chalk scarp at Westbury in Wiltshire.

west, while south-flowing rivers blanketed much of the 'English' chalk in alluvial deposits. In due course, much of the old chalk cover was stripped away by erosion, or domed and tilted by the great earth movements which accompanied the formation of the Alps. Where the eroded edge of a tilted chalk bed meets the surface, a steep scarp face will normally be seen, while behind the face, the escarpment will slope down gently in a dip slope which eventually disappears beneath younger beds of sand or clay. Cycling over chalk scarp and dip-slope country can be an exhilarating or frustrating experience, the slow haul along the sections of road which wind up the scarp faces alternating with long, straight, downhill runs. Although the scarps can be steep and imposing, chalk is too soft and homogenous to produce craggy rock outcrops, and behind the scarps the downland scenery is generally rounded and rolling. Slight differences in the relative hardness of chalk beds may give rise to miniature scarps and dip slopes within the chalk.

The downlands are frequently corrugated by the characteristic dry valleys. Chalk is a porous rock, and rainwater will seep through the beds until it encounters an impermeable rock or reaches the level of the underground water table, when it will emerge in a bubbling spring. The dry valleys, or 'combes', of the chalklands clearly must have formed when factors affecting the water table were different. One attractive theory suggests that they formed when England was experiencing glacial or periglacial conditions. When permafrost conditions prevailed, the frozen subsoil would be rendered impermeable, with the pores and crevices sealed by ice. In such conditions, meltwater could flow freely across the chalk surface to gouge out the now-dry valleys. Some of these combes are permanently dry, while others support small streams or 'bournes' when the water table rises in winter.

The chalk downs support a distinctive assemblage of flora and fauna. In parts of the Chilterns and North Downs, the higher parts of the escarpments are thinly carpeted with the remains of younger overlying rocks. This coating of clay-with-flints produces a sticky, acid soil, supporting quite heavy woodland which contrasts with the surrounding open chalk country. Before the dawn of agriculture, the chalk too, was probably wooded, but the stands of beech which punctuate the downlands do not seem to represent the surviving vestiges of the primeval wildwood. Many, like the lovely beechwoods on the Gog Magog hills south of Cambridge at Wandlebury, were planted by eighteenth-century landlords for landscaping and game-cover purposes.

From the point of view of the naturalist, the downland pasture is one of the most attractive and threatened of our man-made environments. In order to be maintained, the downs must be closely-grazed by sheep or rabbits to produce the short, fine, springy turf which is a mosaic of dozens of small and tightly packed grasses, flowering plants, mosses and lichens. Towards the end of the last century, sheep farming underwent an economic decline. But in many places, a swelling rabbit population perpetuated the close-grazing role of the sheep – until the rabbits' numbers were drastically reduced following the arrival of myxomatosis in 1954. Many of the downlands which have not surrendered to the plough are now grazed by cattle; and since cattle do not browse as closely as sheep, coarse grasses and scrub beech are again advancing. If all human interference were suspended, the downs would gradually return to deciduous woodland, which is the natural climax vegetation. The short, springy turf of old downland pasture is ideally suited to the training of racehorses. Newmarket Heath, the Epsom and Berkshire Downs in England and the limestone pastures of Ireland have several renowned stables. Perhaps the best place in Britain to see typical traditional downland scenery surviving today is the Queen Elizabeth Country Park, which lies to the south of Petersfield in Hampshire.

Like other limestone uplands, the downlands are dry and exposed, so that settlement has traditionally tended to gravitate towards the scarp foot valleys at spring line locations where water is available. Each medieval village tended to house a partially self-sufficient community, and so it would have sought access to a wide range of environmental assets – the common graz-

Beds of tougher chalk were quarried for clunch, a soft building stone which could be carved to produce very detailed work for display in the interiors of churches. These old clunch quarries overlook the snow-covered slopes at Orwell in Cambridgeshire.

The chalk country is poor in building stones and churches may be of flint, which can be knapped (left), or knapped and squared (right).

ings of the chalk escarpment, the ploughlands of the alluvial fans and river terraces below the scarp, and the moist hay meadows and fishing and fowling resources of the clay vales. Consequently, ribbon-like parishes developed, stretching from a narrow river frontage up to or beyond the crown of the scarp. Whether these parishes belong to the historical period or are derived from the ancient estates of prehistory is a matter of debate. These parish

'Glacial erratic' stones – brought by the glaciers from elsewhere – and river cobbles have been used in the construction of the church at Grantchester in Cambridgeshire. Like many other medieval churches, this one clearly incorporated masonry salvaged from an earlier (Norman) church.

forms are most clearly developed in the Lincolnshire limestone scarplands; the ritual study of parish boundaries on the OS Lincoln and Grantham sheet will be familiar to most survivors of O Level geography, and the regularity of such patterns makes one wonder whether a form of planning was involved.

While a layer of chalk rubble frequently provided the footings for the flimsy timber, wattle and daub hovels of the medieval peasants, most of the older vernacular architecture of the chalklands is constructed in timber-framing, old brick, or a combination of the two. Occasionally, one may see cottages which are built of hard, white chalk, or 'clunch', but this stone is more commonly seen today in the patched walls of old barns and outhouses. Cottages built of flints gathered from the surrounding fields are a common sight in most chalk countrysides, and most seem to date from the eighteenth or early nineteenth centuries. In parts of the Chilterns, one can see plastered and lime or colour-washed cottages that are built of 'wichert', a sort of cob composed of puddled clay and chalk.

Despite their relative poverty in building materials, the chalklands contain many fine churches, some of which reflect the medieval profitability of the downland flocks. Some of the poorer churches are built entirely of clunch, and their smooth and cool white interiors contrast with the grooved and flaking textures of the outside walls. Some employ the local flint, which is bedded in mortar to display knobbly whitish nodules or knapped glassy faces, while other church builders have preferred to gather river shingle, which can be seen forming some of the walls in the Cambridgeshire churches at Grantchester and Tadlow. In all such cases, the quoins, tracery and arches must be built in a freestone – either one of the tougher clunches, like the products of the Totternhoe quarries of Bedfordshire, or an expensive imported stone. Only the most generously endowed of the chalkland churches are built entirely of imported stone.

PART TWO

STONE AND PREHISTORIC MAN

Introduction

When we refer to the longest division of mankind's prehistoric past as the Stone Age, or speak of some New-Stone-Age colonists of Britain as the Megalithic, or 'Great Stone', people because of their majestic stone tombs and circles, we acknowledge the ancient importance of stone. One could go further, and argue that without the discoveries which resulted from man's experiments in the use of stone, a vital step on the journey to civilization would have been missing. Without stone, man might never have broken the bonds of natural control, but remained like the bears and apes, his destiny determined by climate, season and natural competition.

Popular conceptions of prehistoric man derive from many sources. Sadly, school teachers commonly choose to introduce our ancient forebears as symbols of a primitive backwardness. In their different ways, Hollywood and the strip cartoons have done the same: they present our ancestors as dim-witted savages fumbling aimlessly through life without organization or ingenuity. Yet, monuments to the purposeful and remarkable creativity of prehistoric man are numerous in Britain. Ancient constructions like the New Grange or Maes Howe tombs, or the stone circles like Callanish or Stonehenge, do not testify to the primitiveness of Stone-Age man. Instead, they tell of people who were organized and motivated, capable not only of considerable feats of engineering, but also of planning and managing long-term building operations.

Similarly, the prime tool of prehistoric man, the stone axe, is often used to symbolize the 'primitive' artefact – presumably by people who have never seen, handled or seriously considered these tools. Well before the close of the Old Stone Age, or 'Palaeolithic', period (around 8500 bc)*, craftsmen were produing axes that were balanced, sophisticated and efficient. During

* 'bc' is used to denote carbon 14 dates

the Middle Stone Age, or 'Mesolithic', period which followed, the art of producing tiny flakes of razor-edged flint to tip an arrow or to tooth a saw-like knife blade had been mastered, while in the New Stone Age, or 'Neolithic', period which began in England around 5000 BC, the axes were polished to a silky sheen, and had become artistic creations, and perhaps status symbols, as well as effective implements for the removal of the natural wildwood. Whether swung as an axe, jiggled as a hoe, or used to tip a plough, the stone blade was indispensable to the Neolithic pioneer farmer. It played a leading role in the Neolithic agricultural revolution, which was arguably the most important event in the life of mankind.

The discovery of copper and gold objects along with the characteristic ritual pottery drinking vessels in the graves of settlers known to archaeology as the Beaker people marks the transition – around 2500 BC – from the Age of Stone to that of Bronze. The importance of stone was scarcely diminished during the Bronze Age. In many upland areas of Britain where boulders could be gathered on the hillsides, stone furnished Bronze-Age man with the walls of his round hut, a kit of chopping, grinding, boring and scraping tools, the slabs which might line his burial cist, and the ring of upright boulders which formed his temple.

In the chapters which follow, I will introduce stone as serving three vital roles in prehistoric British life: as a medium for tool-making; for the construction of religious monuments; and for hut and wall building. It would be difficult to exaggerate the importance of stone to ancient man, for it often provided the material basis for the economic, spiritual and domestic spheres of his life. The evidence repeated here may suggest that a re-assessment of the 'primitiveness' of ancient life is needed. Two millennia have passed since the close of the prehistoric period in Britain, and the reader now belongs to an intensely technological society. If we were cast into a wilderness, we would be quite unable to fashion a serviceable stone axe, engineer the construction of a tomb or temple – assembling and erecting boulders weighing several tons – or build a wall which would endure without the aid of mortar. Yet these skills, which our 'primitive' forebears possessed and which we have lost, were essential for the foundation of modern civilization.

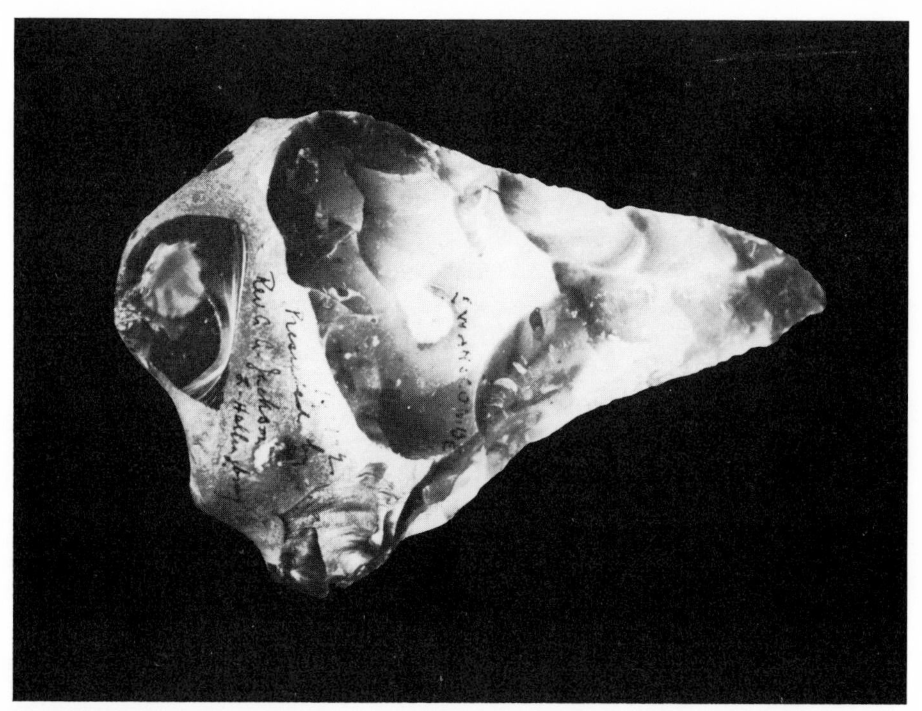

Two Palaeolithic hand axes from the collection in Saffron Walden museum. The one above is from Swanscombe. 250,000 YRS.

Chapter 4

Stone and the Tool-Maker

We will never know the identity of the first British stone tool-maker, nor whether he or she belonged to the mainstream or to an ill-fated backwater of human evolution, nor whether the artefacts produced were truly primitive and inefficient or embodied a measure of craftsmanship. Even so, we can be confident that the first experiments and vital discoveries in the field of tool-making were pioneered far away from Britain, perhaps on the plains of Africa. Neither man nor the making of stone tools was indigenous to Britain. It is important to perceive the development of stone-tool technology within the broad context of human advancement: the first unshaped stone missiles and the razor-edged manufactured hand axes and scrapers of the Old Stone Age endowed our lightly-muscled, fangless and clawless ancestors with the ability to kill, skin and dismember their prey, and thus survive and prosper in competition with other animals. Many thousands of years later, towards the end of the Middle Stone Age, the development of refined axe-making skills allowed the production of tools which could be used to fell the primeval forests, and thus launch the first phase of the landscape transformation which accompanied the introduction of farming – the most significant revolution in man's tenancy of these islands. Until quite recently, the dawn of tool-making was used to define the dawn of humanity. The evolution of man from ape-like ancestors was a gradual process, and scholars sought a criterion which would distinguish man from the remainder of the animal kingdom. The ability to make and use tools seemed to provide the necessary dividing line. However, closer observations of the activities of chimpanzees have shown that this definition of humanity is unsatisfactory. In the 1920s, Professor W. Kohler observed how a captive chimp could sharpen a stick with his teeth and join it to a length of bamboo in order to reach a tempting bunch of bananas which was out of reach. Clearly, the animal was not simply taking advantage of instinct and a handy piece of natural flotsam in the way that a thrush might use a stone as a snail-crushing anvil. He had made a tool to solve a problem in a quite deliberate manner.

More recently, it has been seen that chimps will tear off tree branches for use as clubs or missiles, and in 1960, Jane Goodall saw how wild chimps in Tanzania could delicately fashion 'termite extractors' from grass stems which they then inserted into anthills. Although it can still be argued that there is an important difference between making a handy artefact on the spur of the moment and systematically constructing tools for future use, it now seems wiser to define humanity by reference to the evolution of a large brain capable of more sophisticated and abstract forms of thinking.

Before we look at early experiments in tool-making, let us consider the relevant attributes of the human frame. At first appearance, man seems ill-equipped to prosper in the rough and tumble of the natural world, and his very survival is remarkable. As a hunter or collector of edible roots and berries, he seems unspecialized and inept. He lacks the fleetness of foot of the antelope or zebra, is far less adept at climbing than the ape or monkey, and without fangs or claws is ill-equipped to grapple with a killer carnivore of even half his weight. In the long term, however, this very lack of specialization proved a great advantage, because man learnt to survive in a wide range of quite different environments and embrace many different ways of life. Man's adaptability, combined with a much enlarged brain, almost guaranteed his success, while his third vital asset was his efficient hand. From our distant tree-dwelling ancestors, we have inherited a hand that is long-fingered and pliant, and it was the development of the opposable thumb – the ability to touch the finger tips with the tip of the thumb – which endowed man with the vice-like grip essential in the manufacture of sophisticated tools. *Australopithecus*, the man-like creature who lived in Tanzania and southern Africa some 3 million years or more ago, could and did make tools of stone, and pebbles which have been sharpened by crude chipping have been found in association with his remains. Some of these tools are worn by considerable use, and show that, once manufactured, the chopper was carried and valued. So valuable was the ability to grip and shape a tool that, in the millennia which followed the extinction of *Australopithecus*, evolution ensured the lengthening of the thumb essential to the perfection of the human grip in later hominid species, such as *Homo erectus* and Neanderthal Man.

Bringing our story closer to home, we can be sure that Britain was not a cradle of mankind but, like the other peninsulas of Europe, the accidental destination of migrant bands of hunters, unwitting participants in a northward and westward dispersion of mankind. We know of many British locations where the tools of Earlier Old Stone Age, or 'Lower Palaeolithic', man have been found, yet we still have no clear picture of the ancient communities and their ways of life. There are a number of reasons for this: firstly, the period concerned lies beyond the accurate limits of the radiocarbon clock, which locates most later prehistoric datings in time; secondly, very few of the lower Palaeolithic remains have been found *in situ*, being dredged from disturbed beds of river gravels, culled from the old collections of unscientific antiquarians, or gathered by quarrymen in the course of the destruction of their context; finally, even when the remains are discovered in an undisturbed layer of deposits, they may still be impossible to date with any accuracy, since Lower Palaeolithic geological history is still not properly understood.

There is no doubt that human communities became well established in England in the period following 250000 BC. They lived (as we perhaps do) during an interglacial period which had a duration of some 50,000 or more years. The interglacial climate was probably at least as warm as that of today, and most landscapes in Britain were blanketed in deciduous forest. The era was preceded by what may have been the most severe of British glacial epochs, with glaciers and ice sheets advancing as far southwards as the Thames Valley. Probably the first settlers in Britain arrived well before 250000 BC, during warmer phases, or 'interstadials', of the glaciation. Thick, coarsely-worked hand axes of stone have been found at Kent's Cavern near Torquay and Fordwich in Kent, suggesting settlement in the sunnier southern extremities of Britain during a warmer glacial interlude.

The stone tools which were produced in Britain after about 250000 BC were efficient and workmanlike. Clearly, they had a long pedigree, extending back to genuinely primitive items produced before the emergence of man. *Australopithecus*, as we have seen, shaped crude stone choppers, and for a long time it was thought that primitive eoliths – 'dawn-stones' – had been made in Britain. In 1891, J. Prestwich reported the discovery of such eoliths in surface deposits on the North Downs in Kent, which were then thought to date from before the great Ice Ages. Crudely chipped along their edges, these flints would have taken man's tenancy of Britain back tens of thousands of years into the Tertiary period.

However, modern research has shown that identical flints can be produced by natural processes. When most of Britain lay in the grip of ice sheets, the southern margins experienced a 'periglacial' climate. Here, whilst the soil and subsoil were frozen to depths of several feet, the thin warmth of summer thawed the uppermost layers of soil. Sodden with meltwater, these layers might gradually slither downslope. In the process, flints in these mobile levels would chip against others which were trapped in the permafrost just below. There seems to be no doubt that the Kentish 'eoliths' were produced by this or some other natural process.

The faking of flint tools is also not unknown. The infamous Piltdown hoax of 1908 involved the 'discovery' of the allegedly 500,000 or more year-old skull of 'Dawn Man', which was shown by 1953 to be a fake composed of the jaw of a recently deceased orang-utan and a not-very-old human cranium. In association with the faked skull, and included by the forger to underline its antiquity, were the genuine teeth of pre-glacial elephant and hippo, artificially stained to match with potassium-bichromate, and assorted 'eoliths', falsely provided with an iron patina to give an illusion of great age.

Antiquarian pursuits were popular in the Victorian era. The demand for Stone-Age axes greatly exceeded the supply, and the faking of relics began on a large scale in the middle of the nineteenth-century around certain French quarry sites. Not to be outdone, England produced its own forgers, the most skilful of whom was Edward Simpson, alias Flint Jack or Fossil Willy. One wonders how many cherished private collections still include an unattributed Flint Jack original.

The interglacial period mentioned above is known to scientists as the 'Hoxnian' interglacial, after Hoxne in Suffolk, where remarkable sequences of interglacial deposits have been found in the bed of a former lake which was sealed by later glacial deposits. At Hoxne (pronounced to rhyme with 'oxen') in 1797, the antiquary John Frere noted the presence of some strangely chipped flints lying some feet down in a brick pit. With amazing insight, he recognized them to be the tools of an ancient people who had no knowledge of metal-working. Even so, Frere was not the first Englishman to recognize a prehistoric flint artefact. At the end of the seventeenth century, Conyers, a London chemist, discovered a flint lance-head or axe associated with the bones of an elephant in some London gravels, and recognized the weapon for what it was. But the antiquaries of the next century were cleverer. Obsessed with notions of Britons and Romans, they pronounced that the missile had been hurled by an ancient Briton at one of the elephants brought by Claudius to assist the Roman invasion!

In the course of the Hoxnian interglacial, hunting communities seem to have become quite numerous in southern England, though of course their numbers were closely controlled by those of the beasts upon which they preyed, and modern densities of rural population will not have been approached.

An apparently early Hoxnian community is represented by remains discovered at Clacton-on-Sea in Essex, and known therefore as 'Clactonian'. These people seem to have lived at a time when the environment was heavily forested. Although they lived at an early stage in the settlement of Britain, the Clactonian folk had advanced some way along the road of flint technology. Their tools were not crudely-chipped eoliths, but formed from thick flakes which were struck from large nodules of natural flint. The tool-maker probably sheared off the flakes by rapping the edge of the nodule sharply against a larger stone, which served as an anvil. One side of the nodule and then another would be reduced in this way, and the 'core' which remained might be used as a chopping tool. The serviceable flakes would be gathered, and those with concave edges might be worked to serve as scrapers suitable for shaping wooden spears, while the edges of others were improved for service as skinning and cutting tools. Although the chopper cores and flake cutters and scrapers of the Clactonian people seem less symmetrical and refined than the hand axes of their successors, they embody an accumulating skill and experience. The hand axe, however, became the standard tool of the period. Most were between about three and eight inches in length, and roughly oval in shape. They seem to have been grasped in the hand rather than bound to a haft, and to have served as general purpose tools for severing the flesh or scraping the skin of slain beasts. The more pointed axes will have been useful for piercing the hide in advance of skinning.

Of the remains of people who lived in Britain during the Hoxnian period, probably the most important have been found at Swanscombe, on the south bank of the Thames about eighteen miles to the east of London. When the equable and temperate interglacial era was nearing its close, the debris from various riverside camps was gradually washed along by the ancient Thames to accumulate in gravel beds. Here, in 1935, was found the famous Swanscombe skull. The owner of the skull, who perished more than 200,000 years ago, was almost certainly a very early member of the *Homo sapiens* group and thus closely related to ourselves although whether he was black, white or brown we do not know. Neanderthal Man, on the other hand, is only known in Britain from a few fossilized tooth remains, although the characteristic flint implements of Neanderthal Man are found at several British sites.

The hand axes such as those produced by Swanscombe Man seemed to have had a longer pedigree than the Clactonian tools, being evolved from the crude tools manufactured at the dawn of humanity from convenient pebbles which had been sharpened by chipping. Skilfully formed, with flakes being removed from both faces of a flint nodule, the almond-shaped hand axes testify to the expertise of their makers. It seems that, as the craft developed, a wooden baton or the shaft of a bone was used to rap-off the flint flakes in just the right way. Once the skill had been acquired, the axe swiftly took shape; modern archaeologists who have learned the craft are able to reproduce a hand axe within fifteen minutes.

Settlement in Britain was checked by two more major glaciations before the close of the Paleolithic period. At times, Britain may have been aban-

Sharp flakes of flint which could serve as knives and scrapers were struck from the sides of 'tortoise cores' such as this one, from the Norris Museum in St Ives, Cambridgeshire.

doned to its ice sheets, and Hoxnian levels of population may not have been reached again until the Middle Stone Age or Mesolithic period.

In the course of perhaps 200 millennia – between the end of the Hoxnian interglacial period and the dawn of the Mesolithic era – many different communities of settlers will have visited Britain, and different techniques of tool-making were explored and developed. After about 200,000 years ago, a more sophisticated method of obtaining tools from flint flakes evolved. It was found that if a natural nodule of flint was carefully prepared and shaped into the form of tortoise's shell (a 'tortoise core'), then swift blows to the edge of the core would detach pointed flakes which were instantly serviceable as tools. Large numbers of such cores and flake tools have been found around Northfleet in Kent. Here, just a mile from the older Swanscombe site, a flint workshop must have existed to exploit the rich supply of flint nodules which outcropped in the banks of the old Thames. The Northfleet community were probably visitors to Kent during milder phases of our third major glaciation.

Although the details are still to be discovered, during the many thousands of years which followed we can imagine that groups of hunting folk will have wandered into southern and central England during ameliorations in the bitter glacial climates, to retreat again when more frigid conditions racked the ice- and swamp-girt landscapes. In the later phases of the last glaciation, members of a very mobile and more rapidly progressing Late Old Stone Age or Upper Palaeolithic, culture settled in Britain. Their origins are uncertain, but the waves of settlers who reached the Mediterranean and Atlantic margins of Europe might be traced back to heartlands in the east of the Continent and the Middle East. In the south-west of France, an Eskimo-like culture developed, with members who were skilled in the arts of bone-working, cave painting and flint-working. An offshoot of this people settled in Britain. Their remains have been found at several sites, and the culture is called the 'Creswellian' – after Derbyshire's Creswell Crags, where numerous remains have been found in limestone caves.

While the Creswellian peoples fashioned awls, needles and formidably-barbed spear- and harpoon-heads from bone and antler, the importance of stone tools was undiminished. Predecessors of the Creswell folk had developed a wide range of different tools for specialized tasks, but the most characteristic was the long narrow blade of a razor-like form. These were made with a 'punch' technique: a suitable core of flint was firmly held and long, slender, blade-like flakes were detached from its edges through percussion, using a hammer and a pointed bone or antler tool which was held in the manner of a hammer and punch. The lethal blade which resulted often served as a knife, with one edge razor sharp, the other trimmed and curving over to form a point. A wide selection of smaller flint tools with sharp or chisel-like points or razor edges was manufactured with similar skill, for use in bone-working, skinning, scraping and hole-boring tasks. By the Creswellian period, the technique of 'pressure flaking' – pressing a hard point against the surface and edges of the developing tool – had been learned, and a refined generation of blade tools, including leaf-like spear-heads, resulted.

About 16,000 years ago, an acceleration in the improvement of the climate began to extinguish our last great Ice Age. The Old Stone Age in Britain yielded to the Middle Stone Age, or 'Mesolithic', period around 8500 bc in radiocarbon years (perhaps around 10000 BC in 'real' years, for the radiocarbon dates seem increasingly too young as we go back in time). The warming of the climate was not without occasional chilling setbacks, but hardier species of tree, like the birch, pine and hazel, gradually began to recolonize the British scene and advance northwards. As the bare tundra environment yielded to forest, so the hunting communities were obliged to adjust their way of life to the new conditions. Around 6000 bc, the sea, replenished with waters from the melting northern ice gaps and glaciers, rose to sever Britain from the continent. Thereafter, prospective settlers required canoes.

The Mesolithic peoples of Britain were able to draw upon a tradition of

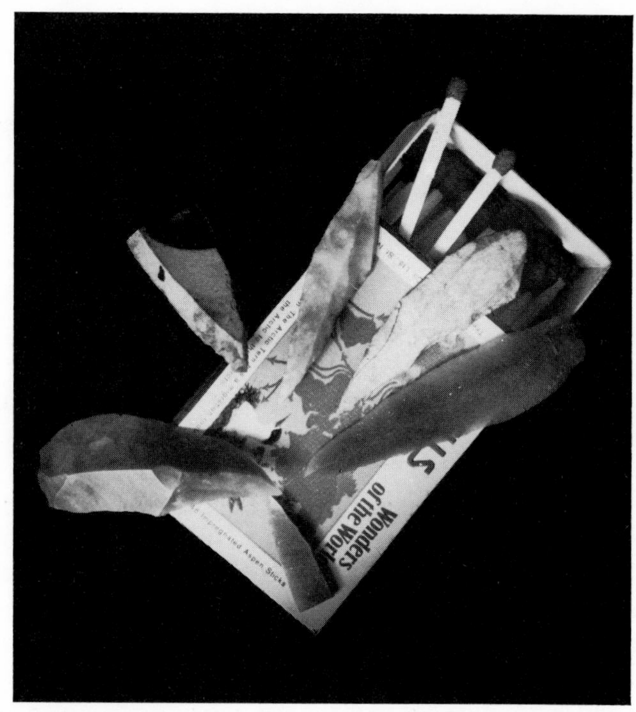

A selection of microliths from the collection in Saffron Walden museum.

skilled craftsmanship in flint which extended backwards over hundreds of thousands of years. Only a very few modern experimental archaeologists are able to reproduce miniature masterpieces in flint to match those of the Mesolithic tool-maker. The changes which were taking place in the environment required changes to be made in the tool-kits carried by prehistoric peoples, and though some larger axes and scrapers continued to be made, the emphasis shifted in the Mesolithic period to the manufacture of minute, precision-made 'microliths' – angular, shaped flakes of flint often much less than an inch in length. These microliths had a variety of uses. Some, with tapering or chisel-shaped tips, served as arrow-heads: the bow was the main hunting weapon of the period. Others were set in line as saw-teeth on wooden shafts and served as cutting tools. Birch bark resin was used to cement the microlith arrowhead or saw-tooth to its shaft. The waterlogged site of a Mesolithic settlement at Star Carr in Yorkshire preserved tightly-rolled strips of birch bark which may have supplied such resin.

The roots of the microlith technique go back beyond the Mesolithic period to the Magdalenian culture, which spawned the Creswellian culture of Britain. Some tiny Magdalenian blades were just half an inch long and one-eighth of an inch in width, and some were set singly or in pairs to barb wooden missiles. The technique of 'pressure flaking' had also been acquired in the Old Stone Age, and was used to shape the cutting edges of tools. Pressure from a small, pointed implement was applied along the edges of the artefact, causing minute flakes of surplus material to split off. The experiences of countless generations of prehistoric hunting folk converged in the flint micro-technology of the various different peoples who hunted, fished and foraged in Britain in the centuries before the introduction of farming heralded the dawn of the New Stone Age.

This saga reveals the curiosity and inventiveness of early man. He did not remain content with one type of tool or technique for very long. He explored different approaches to tool-making and new or better products, became an expert judge of the qualities of different rocks, and adapted his tool-kit and weapons to the slow but profound changes which affected the natural environment. He acquired a range of expert skills which we have long forgotten, and which even the most dedicated of experimental archaeologists now find it difficult to imitate. So let us abandon the myth of the primitive savage, and learn to appreciate the skills and ingenuity of the prehistoric hunter and craftsman, who survived or flourished where we would swiftly perish.

Of the raw materials used by man the tool-maker, flint, for very good reasons, was the predominant choice – the quality of the stone tools produced by Mesolithic communities in the west of Scotland, who lacked access to flint, is often very poor, although at some times flint was imported from Antrim. Flint is not the only rock which responds to expert knapping techniques, and in the parts of the Old and New Worlds where it was available, volcanic glass, or obsidian – black as pitch and glass-like in texture – was frequently worked. Nevertheless, flint is the ideal medium for tool-making. It is hard, with a very fine crystal structure allowing the production of a razor-like edge, which cannot be achieved in coarse-grained rocks. Despite its hardness, flint will fracture in a number of distinctive ways, so that the expert knapper can achieve the controlled removal of a chosen flake simply by delivering a firm but perfectly-directed rap with a baton of wood or bone. It was this suitability for swift and effortless working which particularly endeared flint to prehistoric man.

Fortunately, flint is widely distributed in the British Isles, and is particularly accessible in the English lowlands. Where flint could not be obtained and chert, a close cousin, was not available, men sought other fine-grained rocks like quartzite or basalt, which ranked as very poor seconds to flint. In spite of much learned research, the precise origin of flint is still debated, but it seems that the rock is composed of two forms of silica derived from the glassy skeletons of sponges. The chalk and flints of the English downlands date from the Cretaceous period, around 100 million years ago. At this time, much of what is now England existed as a warm, shallow sea, and the

Unfinished 'rough-outs' for Neolithic polished flint axes from the Norris Museum.

sponges shared the balmy waters with multitudes of minute algae, whose spherical shells of calcium carbonate now form the chalklands. In some uncertain manner, as the sea-floor oozes of shell debris hardened into rocks, acids dissolved the sponge skeletons and the silica solution solidified to form bands of weirdly shaped nodules within the enveloping chalk.

With its glassy, lustrous surface, flint has a visual as well as functional appeal. The flint-walled churches of East Anglia seem to mirror the elements, appearing dark and forbidding on a gloomy day, but glistening lightly in the sun. Once the crusty white coating, or 'cortex', of a nodule is fractured, flint changes colour on exposure to the soil or atmosphere; in addition, regional variations in tint are found. The south-east flints are blacker than the silvery Lincolnshire products or the golden banded nodules of the Swanscombe area. Now, skilled petrologists are learning to identify the origins of particular flint tools.

Wherever it occurred, flint was easily obtained: nodules could be gathered from the surface of the ground, where they endured long after the weathering away of their softer chalk matrix, or prised from nodule bands exposed in the banks of streams. It is not until the New Stone Age, or Neolithic period, that we find evidence of the systematic quarrying of flint and of other useful rocks. During the Neolithic period, the demand for tool-making products must have grown considerably, while the craftsmen were also becoming more selective in their choice of materials. In this era, it becomes quite reasonable to speak of flint industries, factories and workshops, whose products were trade goods for export, often across considerable distances.

The reasons for the buoyant market for flint and other stone tools are not hard to find. The dawn of the New Stone Age, about 5000 BC, was probably heralded by the arrival in Britain of newcomers who brought with them a knowledge of cultivation. While clearings which served as open hunting ranges or pastures for semi-domesticated deer may have been made towards the end of the Mesolithic period, the adoption of an agricultural lifestyle launched a wholesale assault on the primeval wildwood. Before the Neolithic period had run its course, the greater portion of the English lowland forest of lime, elm, oak, ash and alder had been removed, and most of the remaining forest was cleared in the course of the Bronze Age – though in the Scottish Highlands, vast expanses of pine survived into the Dark Ages.

Stone axes were needed in large numbers for the clearance of the native forests; and more were then required as hoes or picks for working the newly-cleared soils. Meanwhile, the adoption of mixed farming and the declining importance of hunting allowed much heavier densities of population to be supported, greatly expanding the market for stone tools. Being brittle, the flint axe is only a second-rate tool when it is swung from the shoulder in the manner of the modern steel axe. But employed either as a wedge to splinter the trunks of trees, or swung from the wrist with a chipping action, it will fell softwood timber with swift efficiency – as experimental comparisons with steel axes have shown.

Well organised and highly productive stone axe factories existed at a number of British locations, while improvements in axe-making technology allowed other hard stones to compete with flint. The finest Neolithic axes have a smooth and polished surface, which was probably achieved by exhaustive grinding against a sandstone slab. The grinding will have allowed a sharp edge to be given to hard stones which refused to flake or fracture in the manner of flint, but which were less liable to break during use. The streamlining of the surfaces of the axes must have improved their performance in the felling of trees, but the polished axe was probably also a considerable status symbol. Cornish greenstones could be polished to a particularly fine lustre; and axes were exported throughout much of Wessex and the south-west, where they seem to have been prized as ritual rather than utilitarian goods. For many years, archaeologists wondered how Stone Age people lacking in metal tools could remove hard chunks of stone like nephrite, steatite and greenstone, or diorite, from an unshaped boulder or rockface. Experiments show that rocks could be cut with flint flakes, but the work was less tedious when the rock was sawn with a string working to and fro in a channel containing sand as an abrasive.

This sarsen boulder in a burial chamber of the West Kennet tomb in Wiltshire has been fluted as a result of its use as a sharpening stone for flint axes.

Many Neolithic axe factories surely remain to be discovered, but we know of a good number of British examples. The case of the underground axe quarry at Mynydd Rhiw on the Lleyn Peninsula of north-west Wales demonstrates the axe-maker's selectivity and geographical knowledge. The bedrock here is a shale, which is quite unsuitable for tool production, but at Mynydd Rhiw, this shale has been penetrated along a bedding plane by a volcanic intrusion of dolerite. Now, the dolerite could have been quarried for the manufacture of quite serviceable axes, but instead the miners exploited the narrow contact zone between the dolerite and the shale. The intense heat which was generated by the molten dolerite caused the adjacent shale to re-crystallize as a tough metamorphic rock which fractures in a flint-like manner. Dolerite sheets slant downwards into the earth at an angle of around thirty degrees to the surface at Mynydd Rhiw, and the Neolithic quarrymen removed the axe material of the contact zone to depths of twelve or more feet. The porcellanite rock which outcrops on the slopes of Tievebulliagh Mountain near the northern coast of County Antrim was formed in a similar manner. Craftsmen squatting on the scree slopes roughed out porcellanite axes, some of which were destined for distant trading; they must have been prized, since Tievebulliagh axes have been found as far afield as the north-east of Scotland and the Thames Valley.

The important factory at Great Langdale in the Lake District, on the lower slopes beneath the peak of Pike of Stickle, has already been described (pp 29–31). Here, craftsmen roughed out the outlines of the axes which were then probably taken to the Cumbrian coast, where gritty New Red Sandstone outcrops were available for the grinding process, and beach sands could be used for polishing. At Ehenside Tarn near Beckermet in the west of Cumberland, such a finishing site was discovered, with sandy and gritty grinding stones, axe-heads in various stages of production, and a polished

Neolithic polished stone axes in various stones from the Norris Museum. From left to right: an unidentified grey stone, possibly Lakeland tuff; greenstone; granite.

axe complete with its haft of beech. Whether the axes were traded by land or by sea is uncertain, though numerous Great Langdale products have been found in Hampshire and the upper Thames Valley. With several other Lakeland factories known or suspected, the area will have been, if only seasonally, a hive of Neolithic industry.

Between Shetland and Cornwall, around thirty non-flint stone axe-making factories are known, or inferred from the examination of finished axes, while in addition, around twenty flint axe factory sites are known. When mapped, these line up in a 'V' formation, following the outcropping chalk westwards from Sussex, and then into Norfolk. Grimes Graves in Norfolk is easily the most impressive of them all. It may have been that until a little after 3000 BC the farmers in and around the East Anglian Brecklands relied in the main upon the tough products of the northern and western axe factories, finding their local flint too brittle for the stern task of tree felling. At Grimes Graves, flint nodules occur in three distinct bands. The two uppermost layers, the topstone and the wallstone, have been weakened by the freeze-and-thaw stresses of former periglacial climates; but a third layer, the floorstone, is unaffected and superbly hard. In certain parts of the thirty-four-acre site where the floorstone outcrops in valley sides, it was extracted by open-cast mining methods; in other places, it lies but a dozen or so feet below the surface, and can be dug from simple pits. Elsewhere, the floorstone is at a depth of twenty to forty feet, and here, remarkable shaft and gallery mines were sunk. It seems that, in all, a total of around 800 pits and shafts were dug at Grimes Graves, and today the landscape is completely pock-marked, like a cratered battlefield.

The shaft mines provide a wonderful testimony to the determination and ability of the miners. Imagine a rimless cart wheel set on end with the axle pointing skywards; the axle represents a shaft, perhaps fifteen feet wide and more than thirty feet deep, while the spokes, seven or more in number, represent horizontal galleries radiating outwards to exploit the floorstone. In the early stages of shaft excavation, wooden shovels may have been used to remove the surface layers of sand, boulder clay and rotted chalk. As the chalk became fresher and tougher, the miners will have resorted to picks made from the antlers of red deer, which were hammered into cracks and jerked to dislodge blocks of chalk, while stone axes will have been reserved for use on the most unyielding chunks. The shaft was sunk until the floorstone was reached, and it was then pursued along the cramped and twisting galleries, where the workers lay on their sides to pick out the flint in the light of a wick burning in grease held in a cup of chalk.

A tree trunk bridged the top of the shaft, ropes were slung over it, and the floorstone and mining dross may have been hauled to the surface in deerskin bags. Then, the nodules were shaped into the rough-outs of long,

One of the galleries from a flint mine at Grimes Graves in Norfolk.

slender, paddle-shaped axe heads by knappers squatting on the waste tips and in the shelters of partly-filled pits, surrounded by a deep litter of waste flakes. The great majority of the Grimes Graves axes were destined for export, and the final flaking and polishing, and the mounting of the head in a cleft pierced through a wooden haft, was left to the importers. Although Grimes Graves was an organized factory, the numbers of miners and knappers employed may not have been large, and the work may have been undertaken during lulls in the farming season. The mines were worked until around 2100 BC, and the occupation of the site continued further into the Bronze Age. The site will reward the visitor: the shaft-cratered landscape is quite remarkable, and it is normally possible to descend a shaft, helmeted, down a vertical steel ladder, to view shaft-foot galleries, which are gated but illuminated.

Another complex of flint mines, some with deep vertical shafts and side galleries, exists at Cissbury in West Sussex, where some of the shafts are cut by the defences of a much younger Iron Age hillfort. Cissbury represents an earlier exercise in the mining method and dates to about 3500 BC. Unfortunately, the shafts here are not accessible to the public. Also in West Sussex, the Blackpatch flint mines share Harrow Hill just north of Clapham with a small, rectangular Iron Age enclosure; shafts and galleries existed here too, but only the pockmarked surface of shaft mining is evident to the visitor.

Many other flint mining sites are known in the southern chalk downlands, and the most important of these must, like the larger stone-axe quarrying sites of the north and west, have been at the centres of far-flung trading networks. Such mines and factories furnished the tools which were essential

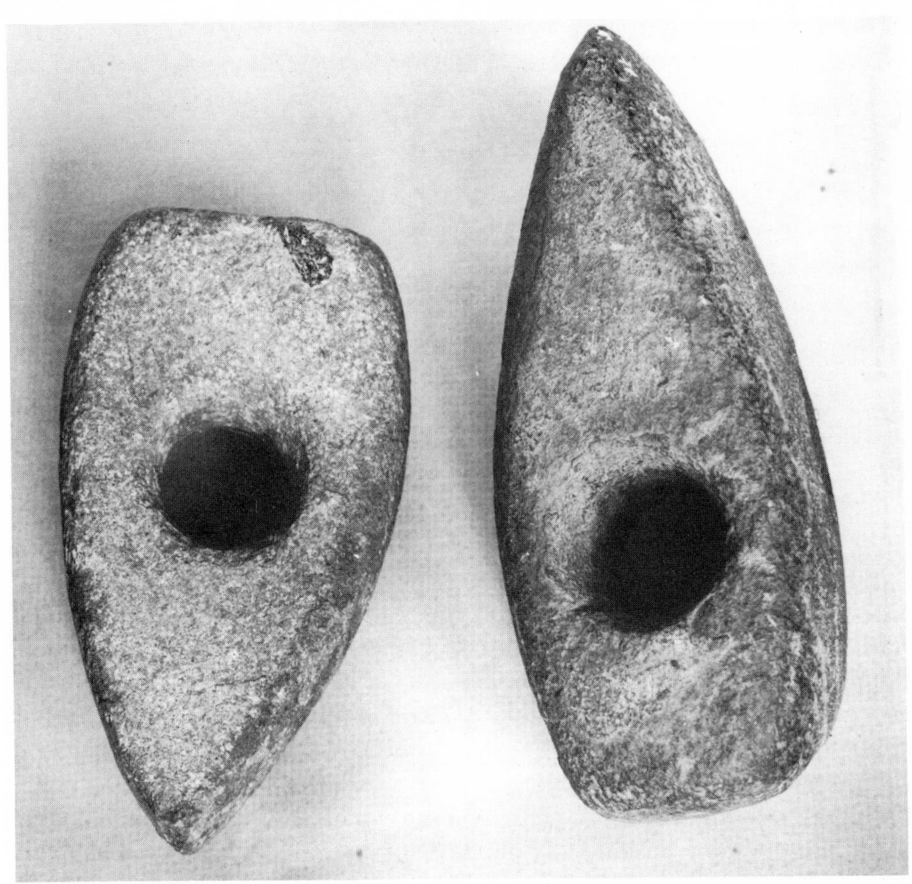

These Bronze-Age maces from the Norris Museum were drilled to take a wooden haft.

to the expansion of farming in Britain, but they were important in other ways. Clearly, New-Stone-Age communities will not have existed as self-contained islands. Albeit indirectly, the axe trade brought the peoples of Wessex and East Anglia, Cumberland and the Thames Valley, and those of many other places, into contact with each other. Thus, it contributed to a broadening awareness of people and places. This process involved the pioneering or development of long distance routeways, some of which, like the Ridgeway and the Icknield Way, can still be recognized, and finds of flint axe rough-outs are common along their looping and branching courses.

The slender, polished blade created by the Neolithic axe-maker was ideal for hafting as an axe or a hoe, and only an expert examination of the tiny scratches on its cutting surface can reveal whether any given blade was employed for felling, hoeing, or as a pick, adze or mattock. Craftsmanship was not confined to the making of superb axe heads, for in the Neolithic period a mastery of the pressure-flaking technique allowed the production of exquisite leaf-shaped or fiercely-barbed arrow-heads, laurel leaf-shaped knives or spear-heads, and even curving, sickle-shaped blades.

Copper-using and Bronze Age settlers in Britain valued their kits of stone tools just as highly as did the earlier Neolithic peoples. Copper and bronze were costly, and their use was largely confined to the manufacture of prestigious ornaments or weapons. The continental 'Beaker' culture, introduced and adopted in Britain after around 2700 BC, was associated with the introduction of copper working about 200 years later. Some craftsmen made flint replicas of the costly copper daggers, and the top people cherished mace-heads of finely polished stone. The maces were laboriously drilled to take a haft by twirling a hardwood stick in a slowly-deepening pocket filled with abrasive sand.

Gradually, more economic methods of bronze-making were perfected, but it was only when the difficult iron-making technology was mastered in Britain after about 650 BC that the new, cheaper and more durable metal began to supplant the highly-evolved and finely-perfected stone tool tradition. Even so, stone querns for grinding grain were still being used surreptitiously by medieval peasants dodging the tolls levied at their lord's mill,

The arrival of metal working did not make flint tools redundant. Here we see an early Bronze Age attempt to reproduce one of the fashionable copper daggers in stone; to the right are two barbed and tanged Bronze-Age arrow-heads in flint – from the Bury St Edmunds museum.

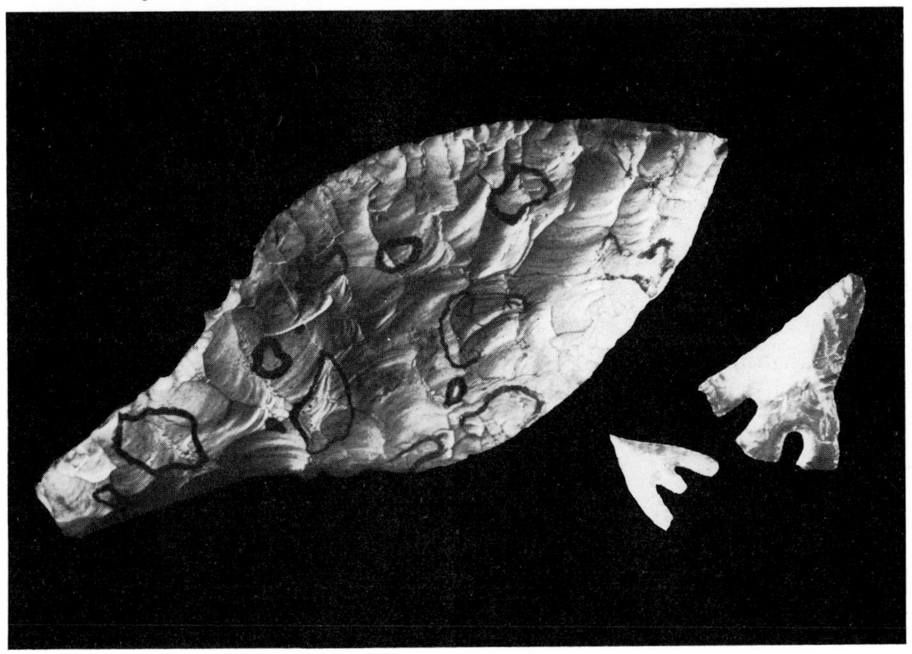

and were employed by Hebridean crofters within living memory. Flint continued to be knapped for use in flintlock armaments or 'strike-a-lights' (the Saxons called them 'firestones'). It was gathered or mined and knapped for use in medieval church building or later house building, and burned and crushed in flint mills to provide a glazing material for the eighteenth-century pottery industry. During the first half of the twentieth century and until quite recently, flint knappers worked at Brandon in Norfolk, producing gunflints for sale to American gun collectors and certain West African warriors who were still armed with flintlocks.

The Brandon industry did not represent a continuation of the Neolithic industry and seems to have developed only in the seventeenth century, following the adoption of the flintlock musket. It did, however, have a number of remarkable parallels. The flint was extracted from floorstone deposits occurring thirty to forty feet below the surface, and it was mined by shaft and gallery methods quite similar to those used by the Neolithic miners at Grimes Graves. Stone was removed using a one-sided pick, which resembled the antler picks of prehistory, and flakes suitable for working into gunflints were struck from a tortoise core. The Brandon workers, however, had access to metal tools and an iron anvil, or 'stake', which rested on a wooden block was used. The quarried flints were dried at the head of the mine, and then the nodules were 'quartered' into small blocks which served as cores. A light iron-flaking hammer was used to dislodge flakes, and these were then knapped into shape using a thin knapping hammer made from a discarded file. The work-rates of the Brandon workers are known, and may throw some light upon the productivity of prehistoric workers in flint. A worker might strike as many as 8,000 flakes a day, and an expert knapper could shape up to 3,000 gunflints. Each pit might have a working life of less than two months; the dross from the industry can be seen in some of the local buildings.

Surveying the ingenuity with which ancient craftsmen prospected for stone, extracted the raw material, and converted it into finely-fashioned generations of tools which were adopted to meet changing needs, we cannot fail to be impressed by the achievements of our distant forebears. A mastery of stone-working was crucial in the human advance towards civilization. It was almost indispensable to the development of agriculture, while the long-distance trading in special axe products will have breached the isolation of many local communities and encouraged the pioneering of ancient routeways. Each new discovery concerning the ways in which stones could be worked carried human progress a step further. If we are to understand the past, we must learn to respect prehistoric peoples, who were creative and imaginative, tougher and no less intelligent than ourselves.

Chapter 5

Caves, Huts and Walls

We have seen that stone provided prehistoric families in Britain not only with the axes, scrapers and missile-heads which formed the main elements of their tool- and weapon-kits, but also with the raw material for day-to-day construction work in the making of dwellings and enclosures. Prehistoric huts and walls of stone in the British Isles are mainly confined to the northern and western uplands, where suitable rock fragments could be gathered with ease. Even in these places, they only survive in the wilder areas which have escaped the plough, or, as in the case of the stone hut village of Skara Brae in Orkney, by a miracle of natural conservation. Nevertheless, in some bare and windswept upland places, the tumbled relics of prehistoric stone-walled dwellings and enclosures are not uncommon, and large expanses of Dartmoor are dotted with the fairy-ring-like debris of prehistoric huts. It has only recently been realized that mile upon mile of ancient field walls also survive upon the moor.

The earliest stone-walled homes in Britain were provided by nature, in the form of caves. Once penetrated, a network of caverns will have offered shelter from the piercing winds and lashing sleet storms which swept across the periglacial landscapes of southern Britain in the dying days of the Old Stone Age. Deep within a cavern labyrinth, the temperature remains constant – cool and clammy, but well above freezing point. However, cave life had its drawbacks, not least in the form of bears and smaller carnivores, which also sought this dark but sheltered lifestyle. Although the most ancient of the Stone Age peoples of Britain are often thought of as 'cavemen', it seems unlikely that cave dwellers were ever a majority of the population. Caves are common only in sea-cliffs and in limestone areas, and elsewhere, people must have sought other abodes, which we know almost nothing about.

Before the amelioration of the climate was sufficiently advanced for trees

A part of the complex of caves known as Thor's Cave in Staffordshire, which was the home of Palaeolithic hunters in the area.

to recolonize southern Britain, hut-framing materials must have been in short supply. Perhaps igloos were built in winter; in other seasons, shallow pits may have been scooped into the ground and covered with tent-like awnings of skin. In the course of the Middle Stone Age, the return of the forests provided a supply of timber for home-making, although very few Mesolithic dwelling sites are known. One apparently popular, if spartan, type of habitation seems to have consisted of a large, boomerang-shaped pit scooped into the ground and covered with a thatch or hides spread over a crude supporting framework of bent saplings. At Thorpe Common in the south of Yorkshire, a few rough stones were gathered together in a crude attempt to improve a natural rock shelter, and this may be an early stage in the evolution of the stone-walled hut.

Of all the numerous British caves which have traces of prehistoric habitation, the most famous must be those of the Mendips. As long ago as 100000 BC, small communities of hunters may have occupied some of the caves which tunnel the grey limestone cliffs like wormholes in weather-bleached oak. Most of the remains date from the final cool phases of the last Ice Age, or from the earlier stages of the Middle Stone Age. The cave known as Hyaena Den, a little to the south of the great cave of Wookey

Hole, may have witnessed many raucous tenancy disputes between hunting groups, who left behind their tools of flint, chert and bone, along with the debris of fires and food bones, and the remains of hyaenas who gnawed upon their prey in the dark recesses of their cavern dens. Flint Jack's Cave, Gough's Cave and Soldier's Hole in the Cheddar Gorge have all yielded flint-tool evidence of cave life. However, the Wookey Hole and Cheddar caves have mostly fallen victim to their spectacular geological endowments: one's enthusiasm for cavemen or stalactites may be severely tested by the crowds, clamour, parking contests and ban on cameras which summer visits entail.

Less besieged is Yorkshire's Victoria Cave, one of the many caverns in the bold limestone landscape between Settle and the swelling bulk of Ingleborough. Here, the oldest remains on soil from the cave floor belong to hyaenas, who were probably responsible for introducing the bones of the hippopotamus, elephant and woolly rhinoceros. Stratified above these are the younger bones of foxes, bears and their prey of red deer and younger still, dating from the last glacial stages and from the Middle Stone Age, are the remains of hunting communities, their flint tools, and a harpoon carved from antler. Perhaps it was the slowly-improving climate which caused the successive hunting bands who visited King Arthur's Cave near Monmouth to gravitate from the black recesses of the two chambers to working and cooking sites in the mouth of the cave. Again, the periodic use of the cave by hunters continued into the Middle Stone Age, and numerous flint and bone tools have been unearthed.

Coastal cliff caves also provided shelter for small, resilient bands of Stone Age hunting groups. Wales has a number of examples, including Bacon Hole in the limestone of the Gower Peninsula. During its Stone Age occupation, the cave did not overlook the sea; for, with so much water locked on land in the form of shrinking glaciers and ice sheets, the sea level was considerably lower than it is today. In the early years of this century, some reddish streaks found on the cave wall were mistaken for Stone Age cave art; and very recently, hopes that Britain had yielded its only authentic example of cave painting, 'discovered' in a Wye Valley cave, were dashed when a panel of experts showed that it was a natural phenomenon. Cave life did not end with the prehistoric era. In the Pennines, some fine baubles dating from the Roman occupation have been found in limestone caverns, including the Victoria Cave. It is most puzzling that the native aristocrats who owned such costly goods were prepared to endure the clammy discomforts of cave life, with puddled floors and constant drips of water issuing unchecked from the cavern roof. Wookey Hole was also used in the Roman period, while in the period between Caesar's forays into Britain and the conquest by Claudius, man-made caverns were occupied near Waddon in Surrey.

The murky solitude of cave life might be welcomed by the prophet, hermit or seer. In Galloway, in a cliff-face about four miles to the west of the ruined medieval kirk (which may stand upon the site of an older chapel set up to commemorate the fourth-century landing place of the Christian missionary St Ninian), is St Ninian's cave. The saint is said to have retired here for prayer and contemplation. True or false, the legend is an old one, and the cave has attracted pilgrims – some of whom carved crosses on the walls and boulders – since the seventh or eighth centuries. Early medieval hermits lived and worshipped in chambers scooped from cliff-faces at Dale in Derbyshire and Red Stone Park in Worcestershire while the river bluff cliff-face at Knaresborough in Yorkshire is a warren of medieval cave dwellings, the most famous being that of the fifteenth-century clairvoyant Mother Shipton.

Among Britain's most recently occupied caves, and certainly those with the most extravagant associations, are the Hell Fire Caves above West Wycombe in Buckinghamshire. Sir Francis Dashwood (1708–81) employed local labour to cut the caves, which burrow into the chalk for a quarter of a mile, in 1750–2. The quarried chalk was incidentally used in laying a road between High and West Wycombe, but the caves are inextricably associated with the debaucheries of Dashwood's 'Brotherhood of St Francis' – later to be known as the Hell Fire Club.

More prosaic are the various medieval quarry workings and man-made caves around the country which were used for storage or wine cellars. Nottingham, situated upon an outcrop of the soft Bunter Sandstone, sits on foundations which are honeycombed by labyrinths of medieval and more recent caves. Some were no more than medieval cesspits, but the odours which issued from beneath old Nottingham were various. The constant underground temperature of 54° Fahrenheit favoured the construction of caverns for use as all-the-year-round maltings, where the malting grain would germinate in winter as well as summer. Nottingham ale was a much sought-after brew. Tanning, another malodorous pursuit, was also practised underground, although most of the caves recently explored by local enthusiasts as part of a rescue operation were post-medieval and used for storage. However, the Director of Excavations has pointed out that the British name for Nottingham was said to be *Tigguocobauc* – 'homestead of caves', and so Dark Age or earlier cavern networks may yet be discovered.

We will never know just where or when the first British man-made, stone-walled dwelling was constructed. Stone-walled huts are known from the Neolithic period, and from the Ages of Bronze and Iron which followed. Their ancient prototypes probably consisted of crude collections of hillside boulder debris, heaped up to break the winds which swept around natural rock shelters. One of the most ancient of the stone-walled settlements is also one of the most remarkable: it is at Skara Brae in Orkney. Although it dates from the later stages of the Neolithic period, and was begun around 3100 BC,

The stone furnishings in the interior of one of the Neolithic dwellings at Skara Brae on Orkney include box-like cots, a 'dresser', a hearth and a sunken box which may have held bait.

it survives almost intact. It is ironic that while not a single peasant hut has survived from the Middle Ages in Britain and the details of these dwellings can only be conjectured from the evidence of their excavated footings, at Skara Brae we have a village which is 4,000 or more years older, yet all but complete.

Skara Brae is special in every sense. Its survival results from its being overwhelmed by storm winds, since when the dunes have moved away and the ancient village has re-emerged. Skara Brae is also special in that it does not represent the typical settlement of its day, but is the product of ingenious and innovative builders who lived in an island setting which might have been designed to encourage the pioneer builder in stone. Though Orkney was only thinly wooded in alder, birch and willow, the Caithness flagstones tempted the Stone Age settlers with slabs of stone, stacked and exposed on the shorelines like planks of wood in a carpenter's shop. The opportunity did not go a-begging: the villagers used the stone not only for the sub-rectangular huts and the alleyways which joined them, but they also collected 'planks' of flagstones to furnish their homes with dressers, cots, hearths and sunken troughs, in which bait or shellfish may have been stored.

All these fittings are still on view, though the roofs of the huts have long since perished, and the thick accumulations of rotting household debris which had enveloped the living village have been lowered and turfed.

This village, however, is not entirely unique. At Knap of Howar, a few miles away on the small but fertile island of Papa Westray, two similar dwellings were excavated among the sandy hillocks. They seem to be a little older than those of the Skara Brae village cluster. The larger dwelling, which measures around thirty feet by fifteen feet within its walls, was connected to the smaller by a low passage. The drystone walls are around five feet thick. The larger dwelling has a paved flagstone floor, with slab partitions sectioning it into two chambers and stone hearths, while the smaller one has several flagstone cupboards and was partitioned into three chambers. Stone also furnished the ancient Orcadian households with tools: the excavators found sharp knives, points and scrapers of flint, heavy, rounded beach pebbles used as pounders, a small polished stone axe, and two stone 'querns', or grinding stones. The most massive of the querns may have been used to crush sea shells to provide grit for the clay used in pot-making. Less completely excavated than Skara Brae or Knap of Howar, Rinyo on Rousay Island is the third member of a remarkable trio of dry-walled Stone Age settlements in Orkney.

The slab-roofed alleyways and the thick and well-made hut walls at Skara Brae reveal a mastery of drystone walling methods. Two building phases are evident at the village, and one may wonder whether the hut-building techniques were first mastered at the sites of older settlements which remain undiscovered. The craft of drystone walling was, in fact, gained before Skara Brae was built, and is displayed in the construction of a number of rather older megalithic tombs, which remain as common features of the Orkney landscape.

Although the later prehistoric ages have left nothing in the way of stone-built dwellings that is quite as complete or remarkable as Skara Brae, and most of the remains lie in remote and barren places, the legacy of abandoned homes from this time is a large one. In the lowlands, the typical hut of the Ages of Bronze and Iron was circular in plan, with low walls of upright posts and wattle work, and a conical roof of thatch. Such perishable dwellings probably had lifespans of little more than a couple of decades, and centuries of ploughing have removed any obvious traces of their presence. (Given the right soil and crop conditions, however, the circular ditches which ringed such huts to catch the rain drips from the thatch may emerge in fairy-ring-like patterns on aerial photographs.) In the uplands, where boulders litter the slopes, the post-and-wattle wall was often replaced by one of stone rubble, often no more than knee-high or waist-high, and this supported a conical roof of thatch, turf or heather. Lying as they do in rocky

Relics of the Bronze-Age village at Grimspound on Dartmoor.

and windswept places which have escaped destructive ploughing, the remains of such dwellings often survive in a recognizable form. Without the aid of excavation, it may not be easy to distinguish between the Bronze Age and Iron Age examples, although in the course of the Roman occupation regular rather than circular hut forms gained in popularity. At some of these hamlet or village sites – like Snowdon Carr on Ilkley Moor – the debris of toppled huts is not easily recognized amongst the natural moorland scatter of boulders, while at others – like Grimspound on Dartmoor – the stone-walled huts, entrances and paddocks are plainly displayed.

One of the most accessible and best preserved of the numerous stone hut settlements is Kestor on Dartmoor, divided by the Teigncombe to Batworthy road, which may well be of Iron Age vintage. It is particularly exciting, because one can see not only the remains of the huts, but also of the hillside patchwork of ancient rectangular fields, the ridge-top droveway used to move the cattle from pasture to water, and an oval compound – the components of an ancient farming way of life. The huts are twenty-five in number, with diameters of about twelve to thirty-five feet. The typical hut had its conical roof supported by a tall central post up to fifteen feet in height, and around this post a ring of shorter posts helped to bear the rafters carrying the covering of turf or thatch. The ends of the rafters were carried upon a rough wall of boulder rubble, gathered from the moor, which stood no more than about five feet above the ground. This was breached by a simple entrance, which normally faced towards the sunny south. One of the best-preserved and largest examples lies in the centre of the oval enclosure; excavation here revealed the remains of a hut-floor hearth, an iron-smelting furnace, and a forging pit.

Even more impressive are the remains of the stone huts in the Anglesey settlement of Din Lligwy. They are built of ancient and massive boulders, which are angular, pitted and contorted. Some of the dwellings are circular, in the old style, others mimic the rectangular buildings of the Romans. The two great circular huts, more than thirty feet in diameter, appear to have been the original dwellings in an undefended British settlement dating from the Roman occupation. Two of the rectangular buildings served as iron-smelting sheds: it is likely that Din Lligwy paid its way by supplying iron to the Roman fortress of Segontium, which dominated the Menai Straits. Towards the end of the fourth century AD, the pentanonal stone wall which rings the settlement was erected, perhaps to afford a little protection against sea raiders from Ireland who harried the coast at this, the time of the Roman withdrawal.

Dozens more ancient hut settlements are known, and many others must still lie undiscovered where peat, heather and bracken shroud the footings of abandoned walls. The craft of drystone walling reached its climax in the construction of the mysterious coastal forts, or 'brochs', of Northern Scotland. Technically they are not prehistoric, since they overlap the early centuries of the Roman occupation in Britain; but they might as well be so, since their builders, and the great seaborne threat which caused the brochs to be built, remain quite mysterious. It seems likely that they were constructed by the local crofting communities, perhaps to meet the threat of Roman slaving. Although almost 600 examples have been discovered, only the broch on Mousa island in Shetland, which towers to over forty-three feet, approaches its original height. Better than the many other partial survivors, it demonstrates their original tapering, slightly bottle-shaped silhouette. Experts are baffled by the brochs, not least because of the remarkable sophistication of their wall construction and design. They seem to have appeared in a developed form which embodies an amazing expertise, and the search for possible defensive prototypes has been inconclusive and controversial. Recent excavations at the broch site known as The Howe on Orkney, under the direction of J. W. Hedges, have produced some surprises. The Carbon 14 date of about 500 bc is centuries earlier than one would have expected, and it is argued that brochs were the defensive nuclei of fortified villages. Perhaps The Howe was an early broch prototype.

The immaculate drystone wall of the typical broch might be some fifteen feet in thickness at its base, but above the reach of attackers it was built in a double form, with the broad cavity containing staircases and chambers.

A section of drystone walling in the entrance to Dun Troddan broch near Glenelg.

Even if Mousa was an exceptionally large example and the typical broch rose to only twenty or thirty feet, the construction by small coastal and island communities of so many of these citadels – most, if not all, apparently in the early centuries of the Roman occupation of England – was an outstanding achievement. It demonstrates that competence in the most demanding aspects of drystone walling was common amongst the northern folk – although the possibility of broch construction by itinerant builders from England has also been raised.

The ability to construct a workmanlike drystone wall existed in the stone-strewn uplands from at least the fourth millennium BC. Occasionally, we find readily-datable sections of paddock walling in association with groups of ancient huts. This is the case with the stone enclosure wall which embraces the Bronze-Age village of Grimspound at the Iron-Age and Roman-Iron-Age village of stone huts called Chysauster in Cornwall, where low stone walls mark out the terraced garden plots which lay behind the dwellings, and at the similarly-dated village of Ewe Close near Crosby Ravensworth in Cumbria, where the stone huts are seen lying in two groups amongst

The skilful hollow-wall construction in Dun Telve broch, near Glenelg.

The remains of one of the courtyard houses in the Romano-British village at Chysauster in Cornwall.

their stone-walled paddocks. It is much harder, though, to date a stretch of wall which lies remote from an excavated and dated settlement. Many field walls – mainly those of a straight and unwavering form – date only from the Parliamentary Enclosures of the eighteenth and nineteenth centuries. Others are medieval, and may sometimes be seen to conform to the corduroy patchwork patterns of medieval ridged and furrowed furlong blocks, or to preserve the elongated forms of medieval strips (as around Castleton in Derbyshire, and below Malham Cove in Yorkshire). But many field walls are, to all intents and purposes undatable, and an uncertain proportion of these must have originally been laid out in prehistoric times, to be blown over by gales and rebuilt many times.

Of all the ancient field walls of Britain, the most thoroughly researched and fascinating are the 'reeves' of Dartmoor, which date right back to the Bronze Age. It has long been known that the moor is generously sprinkled with the ruins of ancient huts, and patched and patterned with the traces of ancient systems of small rectangular fields. But it has only recently been realized that the miles of ruined walls which stripe the moor represent a large-scale prehistoric partitioning of the area. Excavation of certain sections

One of the ruined Romano-British dwellings at Carn Euny in Cornwall; in the foreground is the entrance to a mysterious underground passage, or 'fogou', which may have been used for pagan rituals.

of walling by Andrew Fleming and the Dartmoor Reeve Project has shown that at least some of the walls were built upon older 'lynchets', which formed the bank-like boundaries of ancient fields; that some of the reeves were beautifully constructed; and that they seem to have been built by gangs of workers – at one point a 'gang junction' between a finely-made and a less well-built section of drystone walling was uncovered.

Chapter 6

The Great Stone People

About five thousand years before the birth of Christ, small boats, either singly or in little flotillas, began to lurch to their destinations on the shores of Britain. The boats were probably built of hides stretched on timber or wicker frames, and as they ground to rest on the beaches, the families who had navigated them on the threatening waters of the Channel, North Sea or Irish Sea scurried to release the pigs, sheep and cattle lying bound and bemused in their open holds. Carefully shielding their stocks of seed corn from the salt spray, the settlers stumbled stiffly landward, driving their beasts before them. Filled with a sense of relief and chattering excitedly about the new life which lay ahead, these Continental families brought agriculture to a Britain of dense forests and valley-foot marshes.

For several generations, the immigrants were preoccupied with the stern tasks of consolidating and expanding their agricultural footholds in the landscape, felling the wildwood, burning the fallen timber, breaking and hoeing the ash-enriched soils, and preserving their stock from the depredations of wolves, bears – and maybe the non-agricultural natives, too. Perhaps by the time the youngest of the sea-borne settlers had reached grandparenthood, the clearings were beginning to merge into open fields, while the immigrant communities were intermarrying with the indigenous bands of hunter and fisher folk, and passing on their farming lifestyle. It is possible that at first they buried their dead in mortuary houses built of timber, And then, for reasons which we still do not understand, communities throughout Britain began to build gigantic tombs of earth and stone to receive the corpses of certain selected individuals. Scores of these tombs, either soil-stripped and tumbled or in near pristine form, endure to this day. They are among the most haunting of epitaphs to the peoples who enacted the greatest leap forward that these islands have ever witnessed: the Neolithic farming revolution.

The immigrants' prolonged but dramatic transformation of the various environments they found in Britain, and the construction of their daunting monuments – the tombs, causewayed camps, cursus avenues and then the stone circles – ran roughly parallel in time. All these monuments are concerned, at least in part, with ritual and some powerful and enthralling religion. We know next to nothing that is certain about the systems of belief which are enshrined in the tombs and circles, and the lay reader who seeks to understand his or her distant kinfolk must pick a path through the quicksands and mires of sensible and senseless publications on the subject.

I have said that the popular image of prehistoric man as a 'primitive savage' is an undeserved and derogatory one; but equally unacceptable is the view put forward by apostles of 'lunatic fringe' prehistory that our early ancestors were some kind of supermen. The so-called 'Lost Knowledge of the Ancients' surfaces in all manner of nonsensical forms, such as the harnessing of 'Dragon Power' which flows along 'Leylines' (neither of which exist). The notion that prehistoric monuments may have functioned as astronomical computers has its respectable as well as its idiotic adherents, but the idea that human intelligence and the capacity for progress were implanted by spacemen is surely the most contemptuous of all. It is not only baseless, but it denies the genuine achievements of our resilient and imaginative ancestors.

Serious prehistorians must struggle to make the best sense of the facts that are available to them. Very often, these facts are few and far between, and so the older interpretations must be adapted, expanded or discarded as new information becomes available. Until quite recently, the monuments were seen as the creations of a remarkable race of Mediterranean missionaries. In modified forms, this idea still persists with certain scholars.

The grandeur of the British prehistoric edifices, and the evidence of other massive collective tombs in Denmark, Iberia and Brittany, led the last generation of British prehistorians to suppose that their builders, the 'Megalithic' people, must have come from a Mediterranean centre of early civilization, bringing with them a knowledge of monumental masonry. They were seen as the far-ranging apostles of a Mediterranean religion which had the great tomb as a focus. Crete, where collective stone tombs had been constructed after 2700 BC, was the favoured choice as the origin of the Megalithic exodus. It was argued that the idea of building such tombs was diffused to Portugal and then, around 2500 BC, settlers came to Iberia from the Mediterranean, bearing knowledge of metal-working and the means of constructing drystone vaults using the corbelling method. Thence, the religion and the tomb technology were transported to the backwaters of western Europe.

This set of ideas was flawed by two weaknesses which were endemic in

The interior of the chambered tomb at Stoney Littleton in Avon, looking towards the entrance from the terminal burial chambers.

the 'old' archaeology: the notion that Britain and north-western Europe in general were backward and savage places, deriving all their good ideas second or third hand from the Classical world; and secondly, an over-dependence on typologies. Such was the dominance of typological study that, come what may, every product of human creativity had to have a place in a chain of artefacts extending back to a prototype. The prototypes, of course, all tended to lie in the Classical world.

In any event, the idea of stone tombs in Britain having originated in such a Megalithic movement, at least as I have summarized it above, has been knocked firmly on the head by the new dating techniques developed during the last three decades. Firstly, radiocarbon dating showed that the Megalithic tombs of Atlantic Europe were much older than their supposed prototypes. Secondly, the correction, or 'recalibration', of the radiocarbon dates, which was made possible by the tree-ring dating method, showed that the tombs (and all other radiocarbon-dated prehistoric objects) were older still. Some Breton tombs were found to belong to the fifth millennium BC, and one, Kercado, radiocarbon dated to 3800 bc ± 300 years, has a recalibrated date of *c.* 4800 BC. Thus far, the Breton tombs seem to be little older than the others in Atlantic Europe.

Arthur's Stone, Herefordshire. Slabs outline the entrance passage which was originally covered by an earthen mound. Nine wallstones support the twenty-five-ton capstone which covers the burial chamber.

What, then, are we to make of the Megalithic tombs of Britain? If we can no longer believe in the migration of Mediterranean missionaries, it seems almost as hard to accept that, at roughly the same time, the different peoples of Britain, Portugal, Spain, Brittany, Ireland and Denmark should independently arrive at a religion which placed a mighty emphasis on building imposing collective tombs, and on the veneration of the dead. While there are broad similarities between the Megalithic tombs of different Western European regions, there are also considerable local and regional variations. Perhaps we are discovering a diffusion of broadly similar religious ideas which develop in different ways as they filter into the recesses of the Atlantic West. But whether these beliefs were diffused or evolved separately, it is clear that Britain went on to develop a type of temple, in the form of the stone circle, which is almost without parallel on the Continent.

When we look at the Neolithic tombs of Britain, we encounter an almost Heinzian range of varieties of form. Many of the differences must be due to prosaic contrasts in the building materials available in different places, while others may reflect local preferences, and perhaps contrasts in cultural traditions.

One basic division which seems surely to be rooted in local geology rather than in fundamental differences of belief is between the earthen long barrow – a long cairn of stone rubble – and the chambered tomb in which stones and boulders, often including immensely bulky slabs of rock, have been assembled to form one or more burial chambers. Earthen long barrows predominate almost exclusively in the English lowlands, and are concentrated in chalk areas – great clusters being found in Dorset, Hampshire and Wiltshire, and outlying clusters on the chalk of the Lincolnshire scarplands and Yorkshire Wolds. Among the softest of rocks, chalk has no appeal to the monumental mason, and so the Neolithic peoples of the English chalklands buried their favoured dead under long mounds of earth and chalk rubble, which lacked the refined masonry of the chambered tomb. These were sometimes parallel-sided mounds, and sometimes somewhat trapezoidal in plan. The longest example of all, lying within the Iron Age ramparts of Maiden Castle in Dorset, is around 500 yards long, but most were between thirty and sixty yards in length.

Detailed excavations show that the earthen long barrow was often rather more than a heap of earth. Some seem to have been built over wooden structures which served as mortuary houses. Skeletons found within these long barrows are often incomplete or disarticulated, showing that the body had been left to decompose before burial. On average, an earthen long barrow covered the remains of about six individuals. A couple of examples are known that lack all traces of skeletons, and may have been built as cenotaphs of some kind; by contrast, more than fifty corpses were interred in the Fussell's Lodge barrow on Salisbury Plain.

Clearly, far more Neolithic people died than were ever buried in long barrows. We do not know the entrance qualifications, but it seems that the typical earthen long barrow was erected – with an output of labour which was sufficient, on average, to employ six labourers for four months – until a sufficient number of VIP corpses had accumulated in a mortuary house or similar place. By this time, the remains would range from skeletons to the highly repugnant. The disadvantage of the earthen tomb form was that, once built, it could not easily be re-opened, though new burials were sometimes dug down into the mound.

In parts of the north, where it was surely easier to gather heaps of stones and pebbles than to scrape up a mound from the thin local soils, long cairns resembling the long barrows were built. The chambered cairn, with its internal passages and burial chambers, is a much more impressive creation, a triumph of the drystone mason's craft.

It can be argued that the stone chambered tombs of the different British regions evolved from the designs of older wooden or turf-built houses, or from tombs which were much less imposing or permanent. Thus, the Megalithic chambered tomb could be a 'house for the dead' that has been made to endure and impress through the use of massive stone building slabs. When we look at the surviving examples of aboriginal societies, we find that they tend to be divided into clans – groups of people rather larger than the extended family and rather smaller than the tribe – whose members pool their efforts to further the common goal of survival. Britain was doubtless divided into such groups, and so the Megalithic tomb must generally have been built for the dead – or rather, some of the dead – of a local clan. And so the tomb must also have been an important territorial symbol.

In the West Midlands of England, in the region of the Cotswolds and the River Severn and in the south of Wales, where suitable slabby chunks of sarsen stone, tough sandstone or limestone were to hand, most of the chambered tombs mimic the elongated, trapezoidal or wedge-shaped forms of the earthen long barrow. However, they have slab-lined and roofed burial chambers set within the earth, and rubble barrow mounds, which are usually entered via short stone-faced passages. No two tombs in the Cotswold–Severn group are exactly the same, while the three tombs which are arguably the best-restored and most impressive – indeed, breathtaking – are Wayland's Smithy on the Oxfordshire–Berkshire border, West Kennet near Avebury in Wiltshire, and Belas Knap in Gloucestershire.

Wayland's Smithy, named now after a mythical blacksmith, lies in a knot of trees close beside the Ridgeway, one of the major routeways of its time. The wedge-shaped earthen mound of the tomb is 180 feet long and covers an older, oval long barrow which preceded the chambered tomb. The

The entrance to Wayland's Smithy chambered tomb.

broader southern end of the tomb is forty-eight feet wide, and the entrance passage to the interior breaches its centre – between a magnificent façade of four upright sarsen stone slabs which tower above the intruder. Originally, there were six of these sarsens, and a seventh sealed the entrance passage. A lower 'kerb' of sarsen stones traces the outlines of the barrow, while parallel to its sides, about twenty-five feet from either side, are the quarry ditches from which the earth and rubble of the covering mound were dug. A slab-lined passage, with a roof that slopes upwards to a height of six feet, penetrates the barrow to a depth of about twenty feet; around half-way along its length, burial chambers branch off to left and right, so that the entire chamber has a cross-shaped plan. The spaces between the sarsens of the façade and the burial chamber are filled with short sections of drystone walling.

Archaeologists are particularly interested in this chambered tomb, which dates from about 3500 BC, because excavation in the early 1960s showed that it had been built above an older type of barrow, with fourteen burials lying on a sarsen stone pavement beneath a tent-shaped wooden mortuary house and oval earth mound. In this way, Wayland's Smithy helps us to

Mighty sarsen boulders form an imposing façade to the West Kennet tomb.

trace the evolution of the tomb. The lay visitor will surely be impressed by the first sight of the sarsens of the façade as they loom through the dark shadows of the shading trees, and by the near-completeness of the tomb and its cavernous burial chambers.

Even more striking is the West Kennet tomb, which lies a brisk walk to the south of the A4 near Avebury in Wiltshire. No need to provide detailed instructions, for the tomb is signposted from a layby and the approach is heralded by the sight of the incredible man-made mountain of Silbury Hill (possibly, but not yet probably, also a Neolithic tomb). The West Kennet tomb is sited, perhaps for dramatic impact, on the skyline edge of a plateau, and the second stage of the climb from the A4 (a routeway of at least Roman vintage) is uphill. Although the tomb has had an unhappy past, having been pillaged for bones for use in quack medicine in the seventeenth century, and breached by a roadway and messed about by a crude excavation in the nineteenth century, it is now, since its excavation in the mid-1950s, beautifully restored. Only the surrounding kerb of stones has vanished since they were sketched by John Aubrey in the seventeenth century.

With its row of massive sarsen boulders, its elongated wedge-shaped plan, and its stone-built entrance passage, end and side chambers, the tomb resembles Wayland's Smithy in form – except that West Kennet is built on a grander scale and its enveloping mound is some 330 feet in length. It is entered between three great blocking stones, which were used to seal the chamber and which stand on the site of the original small, crescent-shaped forecourt. The majestic sarsen stones were probably hauled from the immediate area of the tomb (some are smoothly worn by the sharpening of stone axes), while the oolitic limestone slabs, which were used for dry-walling in the spaces between the sarsens, were imported over a distance of at least seven miles. Though younger than Wayland's Smithy, West Kennet continued to be used beyond the close of the Neolithic period. Despite the depredations of the old bone-hunters, two excavations have revealed the remains of almost fifty individuals in the large terminal chamber of the tomb and the four smaller flanking chambers.

With its elevated position, its mighty façade of rippled and contorted sarsen slabs, and the cool gloom of its passage and chambers, the West Kennet chambered tomb, once entered, is never forgotten.

The chambered tomb at Belas Knap in Gloucestershire displays a variant on the Cotswold–Severn type. The form of the enclosing earth mound is again wedge-shaped, and as with some other tombs of this type, the broad end of the wedge is cusp-shaped, curving inwards towards what appears to be a passage of drystone walling with its entrance, which is defined by uprights and lintels, sealed by a blocking stone. This, however, is a false entrance, perhaps constructed to foil the efforts of those seeking to disturb the remains of the occupants. The real stone-lined burial chambers, four in number, are set into the long sides and narrow end of the tomb, open and accessible since the excavation and fine restoration of the tomb in 1928–31. The remains of thirty-one individuals were found when the excavators opened the burial chambers, while behind the false entrance masonry were the later burials of five children and a man.

The Belas Knap mound is 170 feet long, and perhaps because it lacks a screen of sarsens, its impact upon the senses is less dramatic. However, the steep uphill walk from the signposted approach on the lane which leads from Charlton Abbots to the A46 is quite lovely. The first stretch runs through woodland, and the second zig-zags along the edges of pastures, offering splendid views of the Cotswolds landscape.

A second distinctive type of chambered tomb is known as a 'portal dolmen'. These are very common in Ireland, quite numerous in Wales, and occur occasionally in England, most notably in Cornwall. When silhouetted against a stormy sky, tombs of this type are among the most haunting sights the British landscape can offer. They seem to be a fairly early form of chambered tomb, simple in form, but demanding considerable engineering abilities in the transport, erection and hoisting of the monstrous stone slabs which form the chamber. Four (or more) great stones were raised to form

Zennor Quoit in Cornwall: granite slabs form a box-like burial chamber; the massive capstone has now slipped and lies at an angle.

a rectangular chamber, while a fifth gigantic slab was erected as a capstone or roof. Often, the frontal pair of supporting stones projects to form a small forecourt. It has been assumed that the stone chambers of this type were originally covered in a domed earthen mound; some certainly were, but others may always have been free-standing. Certainly the bare, exposed stones are far more striking and evocative of the mysteries of death and the afterlife.

Examples of these tombs abound in the rock-strewn uplands of the west of these islands (and in Brittany), either tumbled and jumbled, or with the massive slabs all in place. All the intact portal dolmens are impressive, and so it is not easy to single out suggestions for visits. Trethevy Quoit in Cornwall, near St Cleer and well signposted on the minor roads, is very dramatic, with the drama deriving from the height of the capstone, more than nine feet above the ground and supported by, in this case, some seven towering slabs, the one guarding the entrance being notched to give access to the chamber.

A completely different aura surrounds the Lligwy cromlech in Anglesey. The massive capstone of the ancient, contorted, whitish local rock rests on a circlet of stumpy boulders. The tomb is guarded by railings, so the capstone, a metre thick and said to weigh almost thirty tons, is like a monstrous toad squatting within its cage. In Wales, the portal dolmen is often known as a 'cromlech', while the Trethevy example introduced the equivalent Cornish term 'quoit'. Pentre Ifan cromlech in Dyfed is arguably the best-known prehistoric monument in Wales, with a slender capstone some sixteen feet long delicately poised above four vertical slabs which are around eight feet tall. Although it seems at first to have the classical portal dolmen form, it also encompasses the tumbled remains of a wedge-shaped covering cairn, a crescentic stone façade, and a forecourt which was embraced by the extending horn-like limbs of the cairn. Thus, the Pentre Ifan cromlech seems to have more in common with the court cairns of Northern Ireland and south-west Scotland, described below.

In Ireland, the gaunt and angular dolmen, stark and almost surrealistic in contrast to the soft greens and greys of the slopes and skies, may be the most haunting image to carry away from a countryside which seems more a part of antiquity than of the present. Here, the portal dolmens, and some other mound-stripped Megalithic tombs, were often known as *Altóire Íodlaidhe*, the altars of idols, preserving memories of the pagan past. With an astonishing abundance of these remains – the result of the richness of the Stone-Age occupation of Ireland, the unsuitability of much of the land for arable farming which would have caused the removal of many monuments in the clearance of ploughland, and the general availability of massive

Lligwy cromlech, an unusual tomb on Anglesey with a massive capstone on short, stumpy uprights.

Pentre Ifan cromlech in Dyfed is one of the most impressive of the Welsh chambered tombs.

stones of many kinds – it is impossible to recommend a 'best' example. Certainly one of the most spectactular of the Irish portal dolmens is the larger of a pair of such tombs at Kilclooney More, behind Kilclooney church in County Donegal. The two portal tombs here are set amongst the debris of a long cairn, and the larger has a twenty-foot capstone carried six feet above the ground on a pedestal of vertical stones, which it overlaps like the cap of a gigantic mushroom.

Ireland has other distinctive Megalithic tomb types. One of the oldest of the designs is the 'court tomb', which seems to be related to the long barrows and chambered tombs of southern England. A distinctive feature is the stone-flanked, open forecourt, lying just outside the gallery which penetrates the interior of the tomb. Stone-Age communities in Ireland experimented with this design and many variations can be seen, though the court tombs are almost entirely confined to the northern third of the country. Frequently, these tombs are seen as exposed and jagged collections of stones, following the collapse of the earth and rubble mound which covered the gallery. This has been the fate of the court tomb of Annaghmare, set on a rocky eminence in the Slieve Gullion area of County Armagh, but what remains is still impressive. Massive, angular boulders are set on end to form a horseshoe-

The larger of the two burial chambers in the tomb at Plas Newydd on Anglesey. Lying in the grounds of a National Trust mansion, this is one of the most accessible of the Neolithic tombs.

shaped forecourt, the gaps being filled with neat drystone walling. The gallery is more than twenty feet in length, now open to the skies, and divided into three burial chambers by three pairs of projecting slabs. Although the Irish court tombs in some ways resemble the English tombs which have been mentioned, the funeral rituals, and perhaps the essential religion, too, were different. Their builders placed the remains of only a very few individuals in the tombs, often cremating the bodies beforehand, but including goods for the afterlife, such as pots and leaf-shaped arrow-heads of flint, along with the charred bones.

The end of the Stone Age in Ireland was marked by the introduction of a knowledge of copper-working. A new type of monument appeared, wedge-shaped in both plan and vertical section. The broader, loftier front of the wedge tomb may have a straight façade of often massive stones, breached by a short gallery, which is generally double-walled, and divided by a stone slab separating the main and ante-chambers. Like all the tomb types, wedge tombs vary in size and form, but one of the neatest and most box-like examples can be seen at Poulaphuca in County Clare, where two long slanting sidestones, a backstone, and a larger sealing slab at the front support the

The remains of one of the small and very early passage graves in the cemetery at Carrowmore, near Sligo.

roof slab. More rugged, but still box-like and with its single slender roof slab covering a thirteen-foot gallery, the Srahwee wedge tomb in County Mayo came to be venerated as the holy well of Tobernahaltora.

The passage grave is in many ways the Rolls-Royce of the Megalithic tombs. It is characterized by the stone-lined passageway which culminates in a wider burial chamber lying deep in the heart of a great domed covering mound. The chamber is often roofed by the corbelling method: in the upper levels of the walls, each ascending course of stones projects inwards beyond the course beneath, until the roofspace becomes sufficiently narrow to be bridged by a capstone. It can be argued that the imposing domed earth mounds of the passage graves are needed to provide the weight of down-thrust necessary to keep the corbelling stones in place. Like the other Megalithic tomb forms, the passage graves come in many sizes and several varieties. They are normally found in western and upland Britain, often grouped together in cemeteries, and generally close to the sea. They probably share a common origin in a religious cult, but seem to have been built by local communities rather than by far-ranging colonists.

In Ireland, the two most notable passage grave cemeteries are Carrowmore, just south of Sligo, with the remains of sixty very early passage graves – mostly simple in plan, many denuded, and some threatened by gravel workings – and the Boyne Valley cemetery, with the colossal tombs of Dowth, Knowth and Newgrange, dating from around 3300 BC. In Scotland, only the majesty of the Maes Howe tomb on Orkney can rival the splendour of the Boyne Valley tombs. The three Irish tombs are impressive not only on the grounds of sheer size – the mound at Newgrange was almost fifty feet high and 250 feet broad – and for the mastery of the corbelling technique displayed, but also for the carved decoration in circles, spirals and lozenge forms which enliven the stones. Whether these labyrinths, zig-zags and serpentine motifs are purely decorative or have deeper, symbolic meanings, we do not know, but they inject an element of sophistication and intrigue which is lacking in almost all other Megalithic tombs in Britain.

We can guess that the construction of the Boyne cemetery provided sufficient work to occupy a force of a hundred peasant labourers for almost a quarter of a century. The many impressive feats of engineering involved the importation of selected stones over great distances. One of the three recesses in the Newgrange chamber contains a pair of enormous stone-hewn basins, one of which is of a granite quarried either in the Mountains of Mourne, more than thirty miles away to the north-east, or in the Wicklow Mountains, a similar distance to the south. The façade of Newgrange was embellished with glistening quartz pebbles imported from the Wicklows, or more distant sources in County Tyrone. The covering cairn is of river pebbles, probably brought in enormous quantities from the Boyne Valley nearby. The stones of the kerb and the surrounding circle are local 'greywacke' gritstones and slates, but they include some examples of syenite from County Tyrone, hauled more than eighty miles by ice or man. Sadly, the façade has been reconstructed in a modernistic,

This basin in one of the recesses in the Newgrange burial chamber is of granite which was brought here over a distance of at least thirty miles.

Corbelled roofing exemplified on a magnificent scale at Newgrange tomb.

airport-like style. Nevertheless, it is clear that the original tomb designers were experts in engineering techniques, organization and design, and had as well the technical ability to procure the rocks of their choice.

The Maes Howe tomb in Orkney is massive, even though it is only about half the size of Newgrange. The neat precision of the drystone walling and corbelling of the burial chamber, which uses the ideally suitable local flagstone, represents a level of craftsmanship unique in Stone Age Britain. Added interest derives from the runic Viking inscriptions which can be seen scratched into the walls of the burial chamber. Some of these, dating from around 1150 BC, describe how the tomb-robbers discovered great treasures – and so they imply that the tomb was not sealed until the dawn of an age of copper, gold or bronze. Although Maes Howe is a cathedral among Megalithic tombs, it has, like Newgrange, become a victim of its own appeal. The hurried formality of the swift guided tour, the queues, bustle and the internal steel barriers all diminish its impact. At the much smaller Welsh passage grave of Bryn Celli Ddu on Anglesey, the visitor has time to stop, think and absorb the haunting ethos of the beautifully-restored monument.

The domed mound of Bryn Celli Ddu has recently been partially restored. Excavation shows that the tomb was built upon an older 'henge', a circular earthen religious monument now represented by an enclosing bank and ditch, and which may have been studded with the stones of a circle. This is in contrast to Newgrange, where the ring of twelve stones – the survivors of an original thirty-five – surround the tomb, but belong to a later phase of religious building than the tomb itself. With its henge predecessor, the history of Bryn Celli Ddu ('The Mound of the Dark Grove') is complex. The mound, when built, covered a pit which was found to contain a human ear-bone and traces of wood, and beside the pit stood a stone engraved with a labyrinth of zig-zag markings. The original stone is now displayed in the national museum at Cardiff, and a cast stands in its place, while the reconstructed mound has been shortened to expose the older pit and its flanking stone. The original mound extended to the ditch of the older henge, and was surrounded by a kerb of heavy stones, but at a later stage it was enlarged until it had a diameter of around 150 feet.

The more closely we look at the remarkable British assemblage of Megalithic tombs, the more we become aware of the variations upon the theme of the massive and imposing stone burial chamber. Sometimes the design seems to be the product of regional preferences and ingenuity; at others, as with the passage graves, we seem to be seeing the far-ranging dispersal of one idea, which then becomes adapted according to local interpretations.

Immaculate finish and superb drystone walling are evident in the burial chamber at Maes Howe. Runic inscriptions left by Norse tomb robbers can just be discerned on several of the stones to the left of the recess.

A shark's-tooth pattern of upright slabs forms the façade at the best-preserved of the two Cairnholy tombs in Dumfries and Galloway.

Scotland has a number of Megalithic tomb designs. In addition to the passage graves, which seem to belong to a later phase in the New Stone Age, there are the ring cairns of the Central and Eastern Highlands – great doughnut-shaped piles of rubble and boulders, often set inside stone circles and surrounded by kerbs of massive stone slabs, all enclosing an open central courtyard. In the extreme north of Scotland, Orkney and the Shetland Islands, there are the 'stalled cairns', in which the long passage or burial chamber is divided into a series of bays by projecting pairs of enormous stone slabs. In the case of the enormous stalled cairn of Midhow in Orkney, the chamber has no less than twenty-eight bays. Elsewhere, in the west and south-west of Scotland, there are tombs which in some ways resemble the court tombs of Ireland, but these Clyde tombs are complex, and may have grown from the gradual enlargement of a single, original stone chamber.

While the layman marvels at the brutal power of the stone tomb architecture, archaeologists classify and re-classify the different types of tombs, noting the similarities that link and define the groups, and speculating upon the ways in which the types have been dispersed and adapted. While the layman may wonder how supposedly primitive people could transport gigantic boulders, set them on end to construct a passage or chamber, and raise other massive slabs as capstones and corbelling stones, the experimental archaeologist can help to answer such questions – using wooden rollers or carpets of straw, ropes, and large resources of manpower to move the stones, and sloping earthen ramps or timber cribs to raise them. But nobody can solve the essential mystery of the compelling religion which is expressed in these thousands of imposing monuments.

The many forms of the Megalithic tombs argue for variety in belief and ceremonial rituals. In the case of the court tombs, for example, the ceremonies – which surely included those accompanying the internment of a corpse – must have been performed in the forecourt which stood amongst the rugged slabs at the mouth of the tomb passage. In the passage graves, on the other hand, the rituals would have been enacted by the flickering light and dancing shadows of lamps or torches in the central chamber, deep within the bowels of the burial mound. With all its miraculous new techniques, archaeology can only examine the material remains of a people. It cannot bring to life the fleeting pageant of a religious ritual, or recover the main features of a religion which existed in the mind rather than in matter. Thus, the Megalithic tombs are like failed and derelict theatres. We can study the layout of the theatre in great detail, find scraps of litter which will tell us something about the audience, but without a time machine we will know little about the plays which were performed on stage.

It seems unlikely that the religion was just a simple form of ancestor worship. While societies which can seldom have lived far above the levels of a basic subsistence were prepared to contribute their vital and energetic members for what must often have been lengthy service as tomb-builders, the bones of the people who occupied the houses of the dead were often treated in a cavalier fashion. Archaeologists who have opened Megalithic tombs have often found the skeletons in disarray. In some cases, the bones of the long-departed seem to have been brusquely swept aside to create space for those of new inmates. In others, it has been suggested, the tombs were periodically cleared of old bones and gradually restocked. Old bones also seem to have been removed from tombs for use in the mysterious rituals performed at the causewayed camps.

The notion of a New-Stone-Age 'Earth Mother' religion has received much more attention than the basic facts merit. Various types of fertility symbols have occasionally been recovered from the interiors of Megalithic tombs, while galleries at Grimes Graves have yielded a simple shrine and the chalk-carved figurine of an 'Earth Mother', or, at least, an extremely pregnant lady. The dark and cavernous interiors of the tombs may also reflect a link with the earth, its fertility, or its mythical forces. It may have been that when the Neolithic communities inserted a mass of bones and ritual debris in a cavernous tomb, they were seeking to placate the earth for the goodness consumed by crops and livestock. The tomb was also a prominent symbol of territorial control, guarded by the ancestors of the folk who still toiled on the surrounding lands. Although some of the tombs favour certain orientations (for example, the heights of the stones in the kerbs of ring cairns tend to rise towards the south-west), astronomical alignments do not seem to be particularly important. One intriguing exception might appear at Newgrange, where a stone-lined, box-like opening lies just above the entrance passage in such a position as to allow the sun to illuminate the central chamber with a slender shaft of light at the winter solstice.

Finally, since stone – in vast, undressed slabs which are weather-worn and smooth, rugged and furrowed by natural contortions, or, occasionally, pecked with undeciphered man-made geometry – is the outstanding feature of all the chambered tombs, one must wonder whether stone itself did not have a central position in the Megalithic religion.

The great tomb-building era continued for some twenty-two centuries. Tombs were rebuilt, expanded and adapted to take account of the changing needs, ambitions and beliefs of their makers, and some tombs are composed of several small adjacent tombs which have been combined. Thousands of years later, when prehistorians sought to name the people who they thought, rightly or wrongly, had navigated the threatening northern seas as missionaries to propel Britain on the first stages of a long march towards civilization, the stone tombs and circles remained so impressive that their creators were commemorated as the 'Megalithic', or 'Great Stone', People.

Chapter 7

The Rings of Stone

Those people who died during the New Stone Age might, if they were members of the designated élite, be laid to rest during their passage to the afterlife in a Megalithic tomb. During their lifetimes, many if not all of them will have worshipped their deities at a stone circle. While the stone circles were the temples of the Late Neolithic and Early Bronze Age periods, they embody at least as many mysteries as the chambered tombs. Like them, they preserve in stone the physical expressions of a compelling belief. They also vary greatly in size and form and express, in ways still mysterious, the ritualistic preferences of the different regions and cultures of prehistoric Britain. We are only now beginning to appreciate these subtle differences. More than the tombs, the circles have provided the fodder for a host of the crackpot theories of archaeology's lunatic fringe. Meanwhile, the circles endure as the undeciphered signposts to a lost religion, beautiful to the eye, baffling to the intellect.

The origins and precursors of the stone circles are largely unknown. Whether or not one chooses to see a common religious current as the force which caused the Megalithic tomb to be adopted in the far-flung provinces of Atlantic Europe, the free-standing stone circle (rather than a stone ring set around a tomb) is a magnificently British creation. While Brittany has a couple of atypical examples, Britain was, in prehistoric Europe, the island stronghold of circle worship. In this respect at least, the islands had their own identity, beliefs and rituals.

Currently, some experts see the Megalithic tombs as a ritual translation into the permanent and monumental medium of stone of an older tradition of home-making in wood or stone. In this way, with the development of a religion which placed a heavy emphasis upon the dead or the afterlife, majestic and durable houses for the dead were built. Similarly, it can be argued that the stone circle has evolved from rings of wooden posts which

The Merry Maidens, one of the most complete of the Cornish stone circles.

may have marked out the settings for communal gatherings. The evidence for this is not entirely convincing, and some of the great concentric circles of posts, excavated at places such as Woodhenge and Durrington Walls in Wiltshire, may be interpreted instead as the supports of enormous ritual buildings of timber and thatch.

At its best, this theory has little to tell us about the beliefs and rituals which spurred so many communities to sacrifice the precious resources of time and energy on the construction of these monuments. Of these beliefs, as I have said, we know very little, despite long and earnest enquiries. In the words of Robert Frost:

> We dance round in a ring and suppose,
> But the secret lies in the middle and knows.

I feel that the secret *does* lie in the middle, in the sense that the stones of a circle serve, like the walls of a chancel or the ropes of a boxing ring, to define an inner space where important rituals are enacted. Even so, the stones must be more than mere 'markers of space', for stakes or hurdles of woven branches could perform this task just as well. At the very least, they provided the chosen ritual site with an aura of grandeur and mystique.

Invaluable work in the cataloguing and recording of the many stone circle forms has been accomplished by Aubrey Burl, one of a number of researchers who have noted that several of the excavated prehistoric monuments reveal evidence of earlier pits dug into the ground.* Probably these held ritual offerings, intended to appease and propitiate the gods or forces of nature which controlled the destinies of peasant farmers. The stone circle might, therefore, be seen as the sophisticated successor of the ritual pit site, and likewise a monument which expresses the human bondage and homage to brutal natural forces.

Sometimes, New Stone Age Britain is described in a way which echoes Thomas Hobbes: 'No arts; no letters; no society; and which is worst of all, continual fear and danger of violent death; and the life of man, solitary, poor, nasty, brutish, and short.' But this surely amounts to a slanted and inaccurate view of the past. The average lifespan in any pre-scientific, subsistence farming community is likely to be short, owing to the high levels of infant mortality. Even in the Victorian era, so many of the modern drugs and medications necessary to support children through the hazardous years of infancy were still undiscovered that the rate of child deaths was terribly high. But, having survived childhood, the average prehistoric peasant had a good chance of living to middle age or longer. The circles may have been erected, at least in part, to propitiate a nature deity, and they may have been built by workers coerced, to some extent, by a controlling chieftain or priest figure; but they could *never* have been built by communities poised on the brink of starvation, or living a 'nasty, brutish' life. Instead, they tell us of societies blessed with surpluses of leisure and food. Without these vital assets, circle-making – which must have conscripted the most vigorous members of the community for weeks, months or, in a few cases, years of labour service – could never have been accomplished.

I have said that the stone circles are temples: the sophistication and grandeur of many examples argue that this is so. However, we are unlikely to explain the circles according to a single social function. In any locality, the neighbourhood circle would be one of the most prominent and well known of the man-made features of the landscape. It would be attractive as a meeting place for secular as well as religious activities. Perhaps, like the causewayed camps and the ring-shaped but stoneless henges, it may have been a focus for the trading of useful commodities between neighbouring communities, and possibly a market place visited by far-ranging traders in stone axes. Let us not forget that in the intensely devout medieval period, the parish church was also a place for secular entertainments of a very boisterous nature, with deals being struck in its porch and games played in the churchyard.

Most people in the developed world, and almost everybody in Britain, has heard of Stonehenge. Maybe around half of the people in Britain have heard

The Stone Circles of the British Isles (Yale, 1976)

The circle stones at Boscawen-Un in Cornwall surround a massive central pillar which is now steeply tilted.

of the other great circle complex at Avebury in Wiltshire, while a smaller proportion of tourists and visitors to the countryside might be able to name one or two other major circles, like the Ring of Brodgar, the Rollright Stones, Callanish or Stanton Drew. However, to seek to understand the stone circles of Britain while paying attention only to the grandest, most renowned and best-preserved examples would be like attempting to comprehend the religious buildings of the Middle Ages by visiting only cathedrals and ignoring the thousands of parish churches. Almost a thousand stone circles are recorded in Britain; we do not know how many more have been utterly destroyed, nor how many little circles lie unnoticed beneath the heather or bracken of upland moors. The circles range from the vast perimeter of the Ring of Brodgar on Orkney, through the middle range of monuments like Grey Croft or Shap in Cumbria, down to a host of tiny circles like Cerrig Pryfaid in Caernarvonshire, which has ten stones remaining, half of them under eighteen inches tall. Whilst the great and famous circles must have been used by the communities from miles around and will have drawn on extensive resources of manpower during their construction, the multitude of lesser circles were surely built by and for the local populations.

Most stone circles are now damaged or incomplete. At Mitchell's Fold in Shropshire, only ten of the original sixteen or more members survive. Only the small minority of circles were imposing monuments, and in this larger-than-average example the tallest stone is only 1.85 metres tall.

In recent years, much attention has been given to the precise surveying and measurement of stone circles. The large and sophisticated monuments apart, it might seem that the stones in a circle are set out in a fumbling and often unsuccessful attempt to form a perfect ring. But, on reflection, the circle is the easiest possible form to mark out upon the ground: one needs only a length of twine or rawhide and a central peg to act as a compass point or anchor. Compared to the challenge of moving and erecting these massive stones, the task of setting them in a circle is surely relatively minor. Now it has been suggested that the circle was just the simplest of a series of monument plans which were favoured in different places at different times. Where the makers seem to have 'failed' to achieve a perfect circle, the monument may, in fact, deliberately conform to one of the elliptical, egg-shaped, flattened circle or compound ring stereotypes.

All this leads us, inevitably, to the convolutions of the astronomical argument, where possibility, probability and fantasy intertwine in a tangle of theory. Some of the strands represent the fruits of precise survey by sincere if controversial experts. Others are based on supposition and inference; others still on the uninformed babblings of apostles of 'alternative archaeology'.

It may just be that some stone circles were deliberately constructed to incorporate certain significant astronomical alignments. The best-known example is Stonehenge, which seems to have been built with its central axis aligned to the midsummer solstice. Experts from fields outside archaeology, including Alexander Thom, Professor of Engineering, and the astronomers Fred Hoyle and Gerald Hawkins, have offered astronomical explanations for the form of Stonehenge. Although their interpretations caused a considerable stir when first published – particularly Hawkins' *Stonehenge Decoded* in 1966 – the highly complex explanations of the monument as an astronomical observatory or computer tend to contradict each other. Sometimes, too, they contradict archaeological evidence, or portray the circle as being unnecessarily complicated and intricate for the mathematical tasks which it was supposed to perform. While Stonehenge may contain some simple astronomical alignments, none of them can be *proved* to be intentional; and the very elaborate interpretations have tended to become less rather than more convincing as the years have gone by.

The remarkable stone circle of Callanish on the Isle of Lewis can be used to exemplify some of the problems of astro-archaeology. The monument enjoys a relative obscurity through its location in the Outer Hebrides. No visitor reaches Callanish without expending a considerable effort – but on arrival, he can appreciate the circle in perfect peace. The layout of Callanish is complicated. At the centre of the circle is a towering, slab-like pillar, almost sixteen feet tall and five feet in breadth, but only one foot wide. Arranged around this central stone in the form of a somewhat flattened circle, compact in form and only about thirty-five feet in diameter, are thirteen undressed stone slabs, around ten feet high on average. A stone-flanked avenue, about twenty-seven feet in width and 270 feet in length, runs north-north-east from the circle, until it terminates in a pair of 'blocking' stones, which are set with their broad axes at right angles to the avenue. Single rows, each of four stones, radiate from the circle due west and to the east-north-east, while an apparently incomplete stone-lined avenue runs southwards from the circle. A single stone, stranded about ten feet to the south-west, may have been part of an unfinished or destroyed outer concentric ring of stones. Finally, a passage grave, probably of a later date, was inserted between the central pillar of the circle and its eastern margin, perhaps destroying some features of the circle in the process. A second, poorly-defined cairn may touch the north-eastern perimeter of the circle and the passage grave which it contains.

Clearly, Callanish is an elaborate and complex circle, and the provision

The stone-circle component of the Callanish monument.

of stone rows and avenues has made it particularly attractive to astro-archaeologists. It is noticeable that we seem to interpret the meaning of stone circles according to the interests and preoccupations of our times. Thus, in the seventeenth century, long before the scientific marshalling of archaeological evidence, antiquaries were transfixed by simple visions of ancient Britons, led by Druids, and by Romans: Callanish had no apparent Roman features, so it was conveniently interpreted as a Druid temple. Before the space age of astronomical discovery and nascent science fiction, H. B. Somerville decided, in 1913, that Callanish was a ruined observatory. But in the space age, of course, it had to be an astronomical observatory, or a computer: in 1966, the astronomer Gerald Hawkins saw it as a device for computing the calendar, while in 1967 and 1971, Alexander Thom perceived it as a highly sophisticated observatory of stellar and lunar events.

It seems quite possible that the two avenues, and perhaps the stone rows, too, were intentionally aligned upon something, particularly since they do not intersect within the central circle at right angles, like the limbs of a perfect cross, but stray from this simple geometrical form. One of the weaknesses of the astro-archaeological case in general is that the number of potential solar, lunar or stellar targets is absolutely enormous. Any juxtaposition of stones is likely to be aligned upon some heavenly event of interest, while it is impossible to prove that any specific alignment which has been discovered is other than coincidental. Since Hawkins' analysis was published, Gerald and Margaret Ponting, two schoolteachers resident upon

Lewis, discovered a missing stone buried at the end of the eastern stone row. They also realized that one of the stones which Hawkins had employed in the calculation of his lunar alignments lay prone in 1860, and may have been erected in the wrong place by the archaeologist Pitt Rivers in the 1880s. Moreover, Professor Thom's interpretations were based upon the astronomical events as they would have been observed in the period 2000–1600 BC, because this was the date given to the stone monuments by radiocarbon methods. Now, in fact, the tree-ring recalibration of radiocarbon dates has pushed the date of their erection back by about 1,000 years.

Excavations took place at Callanish in 1980–1 and revealed a long sequence of adaptations. Timber structures had been erected as early as about 3500 BC, and they were then superseded by the central stone circle and monolith. Then the burial cairn was constructed, and subsequently a kerb of stones was erected around the north side of the cairn. The dating of the avenues is uncertain, but thought to be between 1000 and 2000 BC. Long or short intervals of time separate the various constructional phases, and the religious significance of the site may have temporarily declined during the Beaker period, when the area between the stones was ploughed. Since the construction of the monument may span a period of more than 1,000 years, during which the surrounding area was successively occupied by peoples of different cultures, one must wonder how any masterplan of alignments could possibly have endured through the centuries of change.

The Pontings have been particularly interested in the question of intervisibility between the nineteen different Megalithic sites which have been found in the Callanish locality. They found 342 possible sight lines between the nineteen different monuments, and saw that 220 of these lines allowed one site to be seen from another. Different and interesting possible explanations for some of the alignments, however, have been suggested by Aubrey Burl. The north-north-east avenue was never completed, but it seems to be heading for Tob na Faodhail, 'The Bay of the Ford'. Perhaps, then, it is aligned on water, for many legends link stone circles with water. Burl mentions the legends that the Rollright Stones of Oxfordshire wander down to drink at a spring on New Year's Day, while the great slabs of the Stenness group on Orkney are reputed to tear themselves free from their sockets and roll to the sea on the same night.

As has been noted, no astronomical alignments can be proved to be deliberately built into a monument. Let us not forget that the vast majority of stone circles are small enough to sit on a cricket square, or even a tennis court, and are furnished with stubby, knee-high stones, the greater number of which have often been removed in the course of the passing centuries. It would be virtually impossible to take an accurate astronomical observation from a stone-circle 'instrument' such as this. One cannot help but think that if prehistoric man's prime reason for building stone circles involved the creation of an astronomical observatory or computer, then he would not have used massive intractable boulders to mark the sight lines, but straight and lightweight stakes.

It may well be that astronomical sight lines were incorporated in some of the larger circles, but if so, astronomical orientation was surely a secondary function. Medieval churches may well provide an analogy, for it was common for a church to be roughly aligned upon an east-west axis, with a tower at the western end of the nave, and with the burials in the churchyard often loosely aligned with the legs of corpses pointing to the east. Even so, the prime religious functions of the church had little to do with astronomy; a church was no less a church if it lacked an east-west alignment, or a body no less well prepared for the afterlife if facing to the south.

Archaeologists generally have little to say about the most obvious and striking feature of so many stone circles: their breathtaking beauty. Whatever the main intentions of the circle-makers, there seems no reason to suppose that they were less sensitive than ourselves to the pure visual impact of their creations. I have yet to see anything built by man which is more beautiful than the Ring of Brodgar when the stones are side-lit by the late afternoon sun and glow above the heather against a backdrop of emerald hay meadows, steel blue loch and hazy violet skyline. If the scene is to be rivalled, it may only be by the twisted, silvery slabs of Callanish, grouped like the tense distorted figures of an El Greco canvas, or by the approach to Stonehenge, when the dramatic scale of the silhouette is first realized.

We may never know whether the creation of such beauty was an important goal which motivated the builders of the larger stone circles. Even so, one cannot deny that circle-making provided people with the opportunity to create moving and distinctive passages of landscape in which the natural and the contrived were strangely blended. The skyward-rearing slabs seem unnatural and create weird silhouettes when set on end. But they are hardly ever smoothed or dressed, so they reveal the flakes, pits and grooves of natural texture, and their hues may be echoed in the scree slopes or boulder fields of their setting. Several modern sculptors have sought to recreate the surreal juxtaposition of landscape, sky and huge, intrusive boulder (a theme repeated with success in the landscaped approaches to Dyce airport a few miles north of Aberdeen), but the prehistoric forms remain dramatically supreme.

Stone circles appear to be absent from the eastern lowlands of England. Any that may have existed are likely to have been removed by ploughmen long ago. Perhaps the soft sandstone and chalk of the area were deemed unworthy by the circle-makers, but differences of culture might also be involved. There are a few outlying chambered tombs in Kent, but the absence of tombs and circles on the oolitic limestone of Northamptonshire is surprising. Here, the place-name 'Stanion', which may mean 'stone hut', gives only the merest hint of a lost chambered tomb. In most other parts of Britain, circles are to be found. Normally, the local stone has been employed, but enormous efforts were occasionally expended to obtain a chosen stone from elsewhere. In such cases, at least, it seems that stone itself played a significant part in belief and ritual.

At Callanish, the stone is a light grey Lewisian gneiss, one of the very oldest of the British rocks, streaked with quartz and deeply grained by the tensions of ancient traumas and re-crystallizations. The stones may have been dragged from the rock litter of a loch-side ridge just a mile to the north-east of the circle.

The Ring of Brodgar and the Stenness Stones face each other across a narrow, loch-flanked, partly natural causeway. They inhabit an area which abounds in a flagstone of the New Red Sandstone series, which is easily split into sheet-like slabs. At Stenness, only three of an original twelve or thirteen stones survive; but the ruins remain impressive, as much because of the peculiar characteristics of the flagstone as because of the massive size of the slabs. All are more than fifteen feet tall, quite broad, but remarkably slender, one waif-like sheet being less than a foot in thickness. The Brodgar stones are of the same flagstone; changing colour according to the light, they sometimes seem a pinkish buff, but the hues are spangled with white and lemon lichen blotches, and sometimes a furry, blue-green lichen growth. One of the Brodgar stones bears an inscription in Viking runes, and when I last visited the circle in 1980, another had recently been struck and fragmented by lightning.

The first major circle to be encountered south of the Scottish border is Castlerigg, on a spur-top setting garlanded by mountains and commanding fine views of Lakeland. The thirty-eight dark, lumpy stones of various shapes and sizes are set out in an oval ring with a diameter of around 110 feet, while ten stones form an unexplained rectangular setting within the circle, and an outlying stone stands about 100 yards to the south-west. Most of the stones seem to be ice-dumped glacial erratics gathered from the locality, the heaviest being some fifteen tons in weight, while some have been hauled uphill from the west, a feat which may have involved more than a hundred people straining on ropes to bring the slowly ploughing boulders reluctantly to the hillcrest. Far less well known is remote Swinside in the south-west of the Lake District, perhaps the most perfect of all the British circles.

The most attractive circle in the English Midlands is the poorly-understood Rollright Stones on the Warwickshire–Oxfordshire border. It contains some seventy-seven stones, closely packed to form the perimeter of a circle about 100 feet in diameter. However, the original monument may have had only

One of the Caithness flagstone slabs in the Ring of Brodgar stone circle.

a third or a quarter of this number of stones. The tallest of the Rollright Stones stand almost seven feet above the ground, and may be representative of the original members of the circle. However, weathering has shattered or rotted many of them; some which fell have long ago been carted away as building materials, while others – the merest stumps – have been inserted in the gaps between the original stones. Consequently, it is impossible to reconstruct the basic design of the circle. Even so, this is a very attractive monument; the stones of the local limestone shine bright and white when sunlit, but darken with each passing cloud, and then their ghostly, tottering silhouettes find a counterpoint in the spiky outlines of the pines of the flanking shelterbelt. On closer inspection, the surfaces of the stones are seen to be pitted and pockmarked by weathering, and dappled with slowly spreading discs of yellow lichen. The solitary King Stone stands apart to signpost the circle; less than a quarter of a mile to the east is a ruined portal dolmen.

At Stanton Drew in Avon, the devastation of the passing ages is more obvious, though less so is the intended layout of what was a large and complex circle. Here, in fact, we have the remains of three stone circles, the largest central one more than 120 yards in diameter, a 'cove' formed of three massive boulders, and two ruined stone-studded avenues. Unfortunately, the visitor now sees only an apparently formless scatter of great stones; the cove lies in the back garden of a pub, and most of the circle stones are toppled and overgrown. It is hard to imagine that this wreckage was once a circle complex almost comparable to Avebury in scale. Some of the stones are of an oolitic limestone, and must have been hauled from deposits at least three miles away. Larger are the massive blocks of a local sandy breccia, rock formed from the shattered angular fragments of older stones. At Stanton Drew, finely-grooved and pitted, it takes on the tint of congealed blood. The stones are darkly colourful, and seem to exude an air of great age and wisdom.

The two most celebrated and physically impressive stone circles in Britain are Avebury and Stonehenge, both in Wiltshire. Both employ massive boulders of sarsen stone, a type of rock whose origins are scarcely less interesting than the sarsen circles themselves. Sarsen is known to geologists as 'orthoquartzite', and consists of quartz grains cemented in a matrix. It probably originated in the early part of the Tertiary era – under climatic conditions resembling those of the African savannah lands today – as sheets of silica-cemented sands which capped the surface of ancient plateaus, and this notion is supported by the silicified palm tree roots often found in sarsens. In the course of millions of years after the formation of the sarsen capping, the sheets were eroded into separate blocks. Under the periglacial regimes which prevailed in Wessex during the Ice Ages and their aftermath, these fragmented slabs must have slithered downhill like ships on a sea of mud

One of the most complete sections of the outer circle at Avebury, with massive unshapen Sarsen boulders glowing brightly in the sunlight.

to rest in the valley bottoms, where they can sometimes be seen congregated in sarsen streams. Such a stream can be explored at Clatford Bottom, just to the north of the A4 west of Marlborough, while other sarsen sheets, still in their plateau position, are found in the Aston Rowant nature reserve, where the M40 crosses the Chiltern scarp above the B4009. Sarsen is remarkable for its toughness. As well as appealing to the prehistoric circle-makers, it has also furnished stone for the walls of Windsor Castle, and for the kerbstones of London's streets.

Avebury is a vast and elaborate monument, where the once-precipitous earthworks are at least as impressive as the standing stones. As at Stanton Drew, it is difficult to form an overall impression of the temple because the village intrudes across the site, and because the scale is so huge. The major excavation at Avebury took place during the 1930s, well before the era of radiocarbon dating, but it may well be that, unlike Stonehenge, which is a composite monument, Avebury was built in a continuous operation

Stonehenge, Britain's most famous monument.

towards the end of the Stone Age, before 2500 BC. The Kennet Avenue, originally a double row of 100 sarsen pairs running south-eastwards from the henge, may be a few hundred years later. Avebury consists of a great banked and ditched henge earthwork, breached by four entrances, which still carry roads, and it contains three stone circles. An outer circle, originally studded with ninety-eight great sarsens, has its missing members marked by concrete posts. The northern inner circle, with a diameter of 320 feet, had twenty-seven stones, of which only four remain. It contained a stone cove, like a roofless sentry box, formed from three great sarsens, of which two remain. The southern inner circle, with a diameter of 340 feet, was surrounded by thirty-nine stones, of which five survive and four more are signified by concrete markers.

Impressive today, Avebury must have been magnificently imposing in its original form – with its three circles intact, and its enclosing bank rising some fifty-five feet above the trough of the now half-silted ditch in which the dazzling white chalk bedrock was exposed. During the medieval period, deliberate attempts were made to destroy the pagan monument, and it is notable that the church, unlike most of the village, sits outside the circle. Excavation has shown that one medieval attack proved fatal: a barber was crushed by a toppling sarsen and left thus entombed by his companions.

Stonehenge, of course, is unique. Its individuality is expressed in its makers' laborious efforts to pound and smooth the surfaces of the tougher-than-concrete sarsens, their readiness to import other boulders from beyond the Bristol Channel, and their provision of horizontal lintels to bridge the summits of some of the gigantic sarsen uprights. Rearing above the flat desolation of Salisbury Plain, the circle at first sight appears taller, but more compact, than most photographs suggest. While the lesser stone circles can be regarded as the temples of local communities, Stonehenge, Avebury and, to a lesser extent, Stanton Drew, must be regarded as the cathedrals of prehistoric paganism, built by the combined efforts of many communities and revered across the length and breadth of Wessex, and probably far beyond.

Stonehenge was built, improved, refined and extended over many centuries. Work began about 2800 BC, in an area already studded and striped by ritual monuments in timber and earth. The initial effort consisted of constructing a circular banked and ditched henge about 380 feet in diameter, with the main earthbank uncharacteristically sited inside the encircling ditch. The entrance lay to the north-east and the huge, naturally-shaped sarsen boulder of the Heel Stone was placed outside it. Inside the henge, the fifty-six pits – later named the Aubrey Holes after the seventeenth-century discoverer and the antiquarian, John Aubrey (1626–97) – were dug, and each pit was filled quite soon after its excavation, sometimes with the remains of human cremations. This takes us to around 2200 BC, and the remarkable importation of the Welsh blue stones, which were at first erected in the form of a double circle at the heart of the henge, although this task was not completed and the work was dismantled.

The chosen stone was thought to have come from the Prescelly Hills in the north of Pembrokeshire (now part of Dyfed), although recent tests suggest that the traditionally-favoured sites are unlikely and other Welsh

Stonehenge, a closer view.

It was generally believed that the Stonehenge bluestones were brought from outcrops such as this in the Prescelly Hills. However, a recent scientific analysis of the bluestones is casting doubt on this theory and other Welsh sources for the stone must be considered.

sources are being considered. The boulders must have been hauled to the coast or a navigable river, lashed to seaworthy rafts for the Bristol Channel crossing, and then slowly propelled overland to the monument, probably on a conveyer belt of log rollers, with teams hauling on ropes and others continually moving the passed-over rollers back in front of the stones. (Alternatively, the stones could have been hauled along a 'road' of straw.) This was not only a great feat of engineering, but one involving organization, leadership and perhaps diplomacy, for the stones will have passed through many clan or tribal territories.

Clearly, the site of Stonehenge was sufficiently important that the mountain, as it were, had to be moved. Also, we have conclusive proof that certain stones were highly valued. One is left to wonder whether the chosen stone was revered in its own right, or whether the site that it came from had endowed the stones with a special aura. But at the same time, the blue stones are inherently attractive. They are essentially fine-grained dolorites of various volcanic origins, whose blue-grey lustre is a blend of tints from the crystal stipple of feldspar, olivine and augite, along with grey-green or blue-grey igneous 'rhyolites', and volcanic tuffs. The third phase, running from around 2100 BC to perhaps as recently as 1600 BC, had several sub-phases, and there will have been long pauses between building operations. It began with the levelling of the site following which around eighty massive sarsen slabs, some weighing thirty tons or more, were hauled over a distance of twenty-four miles from the Marlborough Downs near Avebury. They were then pounded remorselessly until their surfaces were smoothed, and erected, using earthen ramps or timber cribs and probably shearlegs, to form an outer circle, 100 feet in diameter, and a U-shaped arrangement within. Horizontal sarsen lintels were elevated to bridge the tops of the sarsens, and pin and socket joints were carved to anchor the uprights and lintels. Two sarsens flanked the entrance, and four small ones were set at intervals around the henge bank. One of the entrance sarsens is lost, the other has toppled, and is known misleadingly as the 'Slaughter Stone'.

About 1600 BC, some of the blue stones which had been cast aside were dressed and made ready for erection outside the sarsen circle, and two rings of pits, the Z and Y holes, were dug to receive the remainder. This design, however, was scrapped, and instead the blue stones were erected as a ring between the sarsen circle and horseshoe, and in a new horseshoe within the great trilithons of the U setting. Finally, the now fallen Altar Stone, a monstrous block of sandstone from Pembrokeshire, was set up at the centre of the blue stone horseshoe. This completed a monument which, in its partly ruined form, has probably aroused more wonder than any other British scene. Much later, the earthworks of the avenue were extended to the River Avon.

Many important questions concerning the ancient monuments in stone remain unanswered. Around fifteen years ago, archaeologists believed that the various types of monuments were distinctive: the earthen and chambered barrows were tombs; the circles were temples; the causewayed camps, with their broken ditch and bank enclosures, seemed to be collecting centres for livestock; while the earthwork henges were ritual monuments resembling the circles. Now these distinctions are blurred.

The forecourts displayed at many chambered tombs, as well as archaeological evidence of rituals enacted within some others, suggest that these monuments were not only houses for the dead, but also religious foci. Large quantities of animal bones and also human bones, perhaps extracted from chambered tombs, have been excavated from the ditches of causewayed camps; so, while they may have been meeting and trading places, these camps seem also to have been ritual centres. Some recently excavated examples plainly had defensive characteristics as well. If the stone circles are judged to be ritual centres *because* of the otherwise unnecessarily elaborate nature of the stonework, then what criteria are we to use to show that the earthen henges were religious rather than social or trading centres? Many stone circles have outlying stones, and sometimes these stones, like the King

Stone at Rollright, signpost the stone circle from directions in which it is invisible to the traveller. Travellers who were unfamiliar with the details of local topography might as easily have been traders in stone axes as pilgrims. If the circle was a religious centre, it could also have been a market and place for secular festivities, while a number have been found to contain burials. The more we look at the detail of the ancient constructions, the more the supposedly distinct monuments seem to overlap.

We must also appreciate that the Megalithic monuments span a vast timespan, and during the passing centuries beliefs must have evolved and changed. The oldest chambered tombs in Britain may have been built before 4000 BC, while the temporal midpoint between then and now comes at about the time that the final improvements at Stonehenge were completed. In some cases, the elaboration of stone monuments may represent expansions wrought by the needs of growing populations. In others, they may reflect changes necessitated by evolving beliefs, though occasionally a monument was deliberately wrecked and replaced, as if it offended a new religion. Around 2700 BC, Britain was influenced by the 'Beaker' cult, which spread across Europe and was associated not only with metal-working, but also with new burial customs and the placing of a ritual beaker drinking vessel with the dead. The rite of burial beneath an earthen round barrow superseded the long barrow and chambered tomb rituals. The beliefs of the new and old religions must sometimes have merged, for some Beaker burials were still made in chambered tombs, while the interest in stone circles seems to have heightened. Some experts have suggested that the abandonment and demolition of the original blue stone circles at Stonehenge may have occurred when the Beaker beliefs of the area were, in turn, displaced by a new religious culture.

We can, of course, try to classify the circles according to their sizes and forms. The possible typologies which emerge are at least as confusing as those which derive from a study of the tombs. Regional differences in circle types are very interesting, because they may suggest provincial religious cults with their own variants on the theme of worship at stone circles. One very distinctive circle type is known as the 'recumbent stone circle', and is largely confined to the North East of Scotland, County Cork and County Kerry. There are numerous impressive examples, with large horizontal stones lying recumbent between two taller flanking stones, in the manner of an altar. There are some differences between the Scottish and Irish examples, but both often have the recumbent stone in the south-west quadrant, and may contain scatters of bright quartz pebbles – another stone which often seems to have had religious connections. One is therefore left to wonder whether religious ties bridged the 500 miles dividing the North East of Scotland and western Ireland. Excavations at Scottish recumbent stone circles show that these monuments were associated with cremation burials and were as much tombs as circles.

Thus, we can only speculate upon the religious backgrounds to the ancient stone monuments – and not least, the position of stone itself in the fabric of belief.

A seemingly different type of prehistoric stone monument is represented by the menhirs, solitary vertical pillars of stone. Most are virtually impossible to date, but they seem to span the centuries of the Beaker settlement and Bronze Age. Again, their function is mysterious. Some may have served as territorial markers erected upon the boundaries of tribal homelands, some seem like gigantic tombstones, marking VIP burials of the Beaker period and Bronze Age, while some paired stones may be sights aligned upon important events in the astronomical calendar. A few are clearly the handiwork of later ages, some of them dating from the historical period, and some of these are no more than rubbing-posts for cattle.

The tallest monolith in England stands in the churchyard of Rudston in the Yorkshire Wolds. Clearly, both the stone type and its chosen site were important, for the gritstone pillar, six feet wide and towering almost twenty-six feet above the ground, was hauled southwards for more than ten miles before erection. The Clach an Trushal stone on the Isle of Lewis rises some nineteen feet above ground level and seems to be the most lofty of the many Scottish examples. As the author of a useful Irish field guide points out, menhirs, which are known in Ireland as *galláin*, are so common that they have never been counted.* One of the tallest, around seventeen feet in height, is Doonfeeny in County Mayo. It has been 'Christianized' by the carving of two crosses at its base.

Different from the circle groupings and the menhirs are the clusters of standing stones in non-circular arrangements. Sometimes, these seem to represent magnificent processional avenues, like the one at West Kennet which I have mentioned. In other cases, the spaced alignments of stones may have served to mark mountain trackways. Such an explanation seems to apply to the succession of stones flanking a trackway which runs through Llanbedr village near the Dwyryd estuary in Wales, a track perhaps used by Bronze-Age metal traders. Occasionally, the alignments may mark astronomical orientations. This explanation cannot apply to the remarkable Devil's Arrows on the outskirts of Boroughbridge in Yorkshire. Three of an original four gigantic gritstone pillars still stand here, the tallest towering over twenty-two feet from the ground. They are slightly out of alignment, discounting an astronomical explanation, but as at Rudston both stone and site were important, for the massive slabs were probably quarried at

*Anthony Weir *Early Ireland: A Field Guide* (Blackstaff, 1980)

One of the two great monoliths which form the Pipers in Cornwall.

Knaresborough, more than six miles to the south-west. No less puzzling are the numerous Dartmoor stone rows, like the one at Merrivale. Here, in an area studded with cairns and a small stone circle, there are two double stone rows, one 596 feet in length, the other 865 feet. They are too narrow to be processional avenues, and both terminate in 'blocking stones' which close their ends. Though they cannot compare with the spectacular parallel stone rows of Carnac in Brittany, Britain contains a number of these single or

Two of the three great gritstone monoliths now known as the Devil's Arrows. Notice how several millennia of erosion by rainfall has fluted the tips of the pillars.

double rows, and they are completely puzzling.

Decorative or symbolic carving of an elaborate nature is very rare on the Megalithic monuments of the British Isles, though the Boyne passage graves in Ireland are remarkable for their geometrical carvings. Isolated examples are found in a few British passage graves. Small, cup-shaped hollows, sometimes surrounded by a concentric carved ring or rings, are, however, quite

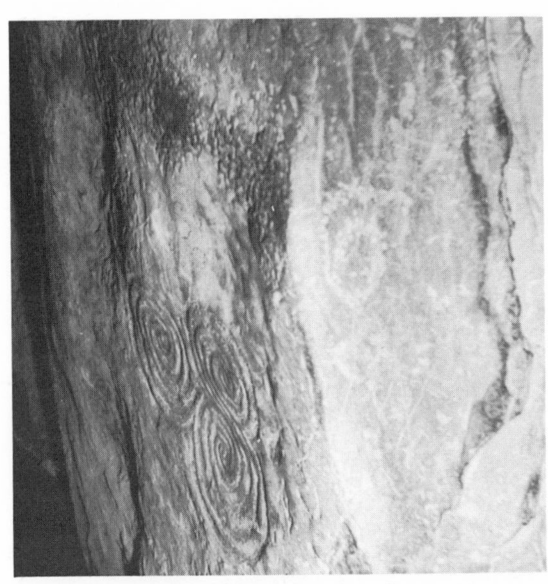

Spiral decorations pecked on the stones in the chamber at Newgrange.

commonly carved both on man-made monuments and on natural boulders. These cup and cup-and-ring markings seem to belong to both the Neolithic and Bronze Age periods, but mainly to the latter. Their significance is quite mysterious, and none of the many explanations offered for them seems particularly convincing. In the 1960s, Ronald Morris collected fifteen different examples of explanations which might be possible, and a further ten which were quite fanciful. In short, this is one of the prehistoric mysteries which may never be solved.

From this brief review of prehistoric man's preoccupation with stone in relation to religious and ritual uses, it is clear that the material, perhaps because of its rugged permanence, and perhaps too because of its intrinsic beauty, aura or even its mystical associations, figured largely in the spiritual side of prehistoric life. Let us not forget, though, that the Neolithic and Beaker peoples were essentially peasant farmers. Agricultural subsistence and expansion will have been their prime concerns, and only a small fraction of their time and labour resources could have been diverted into 'unproductive' monument-making. Many of their stone creations were quite small, and most used accessible outcrops or scatters of local rock. Even so, on occasions the communities were sufficiently organized and motivated to attempt successfully the construction of grandiose and imposing monuments, importing enormous slabs of a chosen stone over distances of many miles. Not until or since the Middle Ages was there a comparable eruption of religious creativity. The sarsen component of Stonehenge alone is believed to embody more than 30 million man hours of labour.

PART THREE

The Rise and Fall of Building in Stone

Crowland Abbey – superb masonry in a fenland area deprived of local resources of building stone.

Introduction

*B*ygone generations of masons have provided Britain with a splendid and varied legacy of buildings in stone. A large proportion of these buildings are objects of beauty quite independently of their historical associations. Each one is a testimony to the craftsmanship of its maker, and to the customs, values and ambitions of its age.

For a society to build in a manner which is magnificent and enduring, the currents of ability and intent must run strong and deep. The circumstances must be right, and the tides of social change must be running strongly behind the visionary dreams. In the centuries which followed the Norman conquest of England, the pieces fell into place. The will, the wealth and the power to build – and to build big – were blossoming. The experiments were made and the craftsmanship accrued. In the period of the High Middle Ages, fired by the energies and ambitions of the social climate, the crucible of creativity overflowed. These times of tyranny, oppression, coercion and war gave birth not only to the cliff-faced and turreted citadels which one might expect, but also to the unprecedented and unsurpassed magnificence of the cathedrals and monasteries of a confident and increasingly opulent church.

Towards the close of the Middle Ages, much of the energy which was needed to sustain the creative momentum became dissipated. Beliefs and visions changed, the church fell victim to its own vanity, the Welsh frontier citadel to its own success, the baronial castle to natural redundancy.

The saga of building in stone might be compared to the life-cycle of an imaginary volcano, with the prevailing social climate representing the source of its life energy, controlling the intensity and nature of the eruptions. After the sporadic simmerings and sputterings of the Roman and Saxon periods, our volcano blows its top in the centuries which follow the Norman capture of England; the eruptions are sustained for a full three centuries,

The adoption of Classical architectural styles provided fewer opportunites for the carvers of stone to display their skills. This Classical façade was grafted on to the medieval village church at Wimpole in Cambridgeshire.

each outburst more forceful and far-reaching than the one which went before.

Gradually, however, the great vent of the volcano becomes choked: the monastic and cathedral-building campaigns begin to lose momentum; the centralization of power under the Tudor monarchs confronts those heirs to the baronial castle-building tradition who have survived the Wars of the Roses with a might which no provincial potentate can dare to affront; and ultimately, the Dissolution of the Monasteries signifies the subsidence of one of the most vital phases in our architectural history.

But the volcano is as yet far from dormant, and an abundance of pent-up energy still blasts its way to the surface through new channels. With the main surges of ecclesiastical building at an end, old beliefs challenged and ancestral strongholds exposed and vulnerable, members of the old aristocracy and of the rising classes of entrepreneurs seek new outlets for their fortunes. Wealth which might once have been invested on a passport to Heaven via a generous ecclesiastical building endowment, or spent on a terrestrial stronghold, is now diverted into the comforts and splendour of the stately home. At first, these mansions ape the moats, turrets, crenellations and gatehouses of the *bona fide* fortresses, but as the Middle Ages are left behind, status is expressed in the different adornments of an unrestrained opulence.

The stately mansion in stone or brick was well established in the British scene by the time members of the intermediate classes of traders, yeomen and successful artisans began to aspire to dwellings of stone. In the cities, the vogue for domestic building in stone was encouraged by the periodic ravages of fires, which might incinerate streets or neighbourhoods of tightly-packed dwellings and workshops constructed of thatch and timber-framing. Several municipal authorities sought to enforce building codes demanding the use of brick or stone.

In the countryside, the vernacular building tradition, even in stone-rich regions, was almost invariably based on one of the numerous variants of the timber-framing technique. The fine legacy of cottages and farmsteads which display the local stone in regions such as the Pennine Dales and Cotswolds accumulated largely in the aftermath of the post-medieval Great Rebuilding. They are symbols of the decay of feudalism, and of an age when those countryfolk who stood a little above the peasant ranks began to aspire to homes which were durable and, while still functional and unadorned, expressive of modest rises in wealth and status. Built by local 'roughmasons', who were gifted only in the basic skills of their craft and might never aspire, like their medieval forebears, to work on majestic cathedrals or fortresses, these stone cottages and farmsteads belong to the last stuttering phases of the great stone building era.

Buildings of stone were never cheap. Under the feudal system of values, only the church, the monarchy and the mighty aristocrat could contemplate projects of incredible cost and scale. In the construction of castles, no expense would be spared nor corners cut to make the citadel equal to all conceivable challenges. In the building of cathedrals, the eventual cost of the undertaking, and the fact that its sponsors might not live to see it completed, were secondary to their aims of glorifying the Creator in the finest manner that craftsmanship and fashion would allow. As feudalism gave way to capitalism, the concepts of cost and cost-effectiveness became dominant instead. The visions of the master mason and the architect were contained within financial restraints. Work had to be accomplished within a given time and for a given cost. The finely-carved details which were the joy of the Decorated style of church building surrendered to the greater simplicity, but still breathtaking sweep and scale, of the Perpendicular style. Then, the Middle Ages past and gone, the spiky elaborations of the Gothic styles were overtaken

The construction of tombs and memorials provided employment for the masons of many ages. In the eighteenth century, the greater monuments tended to become rather theatrical, but those of the seventeenth century are more stark and direct – like this grave slab preserved in the church at Orwell in Cambridgeshire.

by the austerity and restraint of Classical designs. In the construction of palaces and mansions, the wealth and the will to build on an imposing scale persisted long after the capacity to emulate the finely-wrought details of the ecclesiastical buildings of the High Middle Ages had been lost.

Stone buildings continued to be constructed during the nineteenth century: smaller mansions, prestigious town halls for the upstart industrial centres, bridges and new urban churches were still often built of stone – and there was a remarkably widespread endeavour to restore and reconstruct the legacy of decaying medieval churches. Much of what may appear to be medieval church masonry is found, on closer inspection, to be Victorian restoration; but much that was done was over-exuberant, and at odds with our modern ideas of conservation.

Increasingly, however, cost constrained the use of stone. In the Tudor period, brick emerged as an expensive but highly prestigious alternative. In the decades and centuries which followed, the costs of brick production fell and fell, while the costs of quarrying and dressing stone remained high. Transport costs from quarry to building site had always been a major restriction upon the use of stone. By the time the canals and railways were able to provide an effective transport medium and the age of timber-framing was almost past, it was the cheaply-produced brick which carried off the prizes.

The age of stone building is now almost dead and buried, and so many of its splendid creations have been cast down, or cower beneath the gaze of the development profiteer, the muddle-headed council and the misnamed planner. In the middle of this century, city councillors (mainly well-intentioned if naive in their pursuit of the brave new world of housing for all, free-flowing traffic and rational development) listened eagerly to the apostles of the latest architectural trends. The results we all know well. To make room for a tedious succession of lookalike city centres, many fine old buildings in timber, brick and stone were levelled. In their place rose the glass and concrete tower blocks which even the architects now realize were a ghastly mistake. Along came the tatty car park and the tacky shopping precinct and, often worst of all, the cheap brick and concrete parodies of the civic architectural tradition. It is all very nasty, and just as closely in keeping with the spirit of its age as the medieval fortress, abbey or cathedral.

A visit to an old castle, mansion or abbey ruins can be at the same time reassuring and depressing. It may be reassuring to find that there are scores of tourists who are attracted by historic places and by the achievements of past generations of builders. At the same time, the inability of parents

In the eighteenth and nineteenth centuries, stone was widely used in the construction of humbler dwellings, like these cottages at Markington in Yorkshire, which employ the local Magnesian Limestone.

to answer questions about the meaning of the relics and the widespread lack of the knowledge needed to see the relics in any meaningful historical context must detract from the enjoyment of many trips. We have created a technological and materialistic society within which most families enjoy the means to travel and to tour. By and large, however, we have failed to create education systems which fulfil the popular craving for information about our past, our monuments, our heritage. The brighter and more alert a child, the more likely that he or she will be directed in the narrow pursuit of qualifications and 'useful skills'. The difference between the layman and the expert is not that the latter knows all the answers, but that the expert knows the questions to ask: the greater the expertise, the more pertinent the questions. Education in Britain fails because potential enthusiasts are not provided with the background information which is essential before the first, basic questions can be asked. We are urged and trained to work, but it often seems that the enjoyment of leisure is deemed an irrelevant pursuit. In the chapters which follow, some information about the history of British building in stone is given which may help the reader to look at old buildings in a new and questioning way. If I succeed, the pleasure will not come from the reading so much as from the looking and probing which may follow.

Chapter 8

Locating the Stones

Whether in active use or abandoned and overgrown, the quarry is seldom a place of beauty or romance. The visitors who flock to admire the magnificent stone buildings of Britain or pause to view the more homely charms of a stone-built cottage seldom spare a thought for the origins of the raw materials or the techniques which were employed to extract and shape them. Even so, the scores of derelict quarries which now exist only as rolling mounds of dross and debris, or as bush and briar-entangled pits, were much more than *just* the mines of the building industry. Each quarry site witnessed the application of methods of extracting stone which were appropriate to the qualities of the materials involved, and were conditioned by the prevailing technologies. Many were also the places where the hopeful recruits to the ranks of the masons learned the basics of the craft and accumulated their skills. Behind every great stone building there is at least one great quarry: Salisbury Cathedral is built of Chilmark stone, Canterbury Cathedral largely of Caen stone, and Bodiam Castle is of Wealden stone. Most of us know about the buildings, but few remember the stones.

Quarries date from many different ages, come in a wide variety of sizes and forms, and exploit many different types of stone. From the point of view of the landscape historian, it is an unfortunate fact that the more recent quarrying activities have normally tended to obscure the evidence of earlier phases. Consequently, it is often difficult to date the origins of work at most of the older sites. The earliest evidence of systematic and organized stone quarrying operations in Britain comes from the specialized axe-factory workings of the Neolithic and Bronze-Age periods, which have already been described. The purposeful extraction of stone in considerable quantities for building and constructional works must have developed at a much later date.

Diverse stone quarries were established by the Romans in the course of

Barnack rag is a tough, shelly limestone which was quarried in Roman times and was exported throughout East Anglia during the early medieval centuries; here it is used in a wall at Ramsay Abbey.

their efforts to defend, exploit and civilize the British imperial outpost. Quarries were opened to provide the stones for Hadrian's Wall, and at the other end of the colony, the Romans exploited sites on the Isle of Wight, the Kimmeridge shales of Dorset, the high-grade oolites of Ham Hill in Somerset and Bath, and the sandstones of the Bristol area. In a few places, the Roman quarries survive, as near Aldborough in Yorkshire, an important settlement where the Roman quarry which furnished the town was weirdly embellished in the eighteenth century by the addition of niches containing statuary. In other places, quarries which were pioneered by the Romans were greatly expanded in the medieval period, and so the original workings disappeared beneath the debris of later operations. The Barnack quarries, shuttled from county to county and now resting in Cambridgeshire, may be the most important example. Before the Norman conquest, the shell limestone, or 'rag', was being exploited by Bury St Edmunds Abbey, and rights concerning the narrow canals or 'lodes' essential for the export of the stone gave rise to a number of disputes. The Conqueror intervened to prevent the Abbot of Peterborough from interfering with the transport of Barnack rag to water, while in 1192 the monks of Ramsey reluctantly allowed the Sawtry community to transport stone along a lode leading to their abbey. The

Barnack rag was worked out in the fifteenth century, though quarrying continued for a while on other beds. Today, the pits and waste tips of the Roman quarrying era are unrecognizable among the rolling sea of humps and hollows which bear witness to the medieval importance of Barnack village.

Many other Roman quarrying sites were greatly expanded in the medieval era. Stones for specialized uses, like the roofing slates of Collyweston in Northamptonshire and the luxurious fossiliferous 'marble' of Purbeck in Dorset, were first exploited in Roman times. Centuries after they had been opened by the Romans, the Chilmark quarries in Wiltshire yielded the pale buff stones used in the building of Salisbury Cathedral, the Ancaster quarries provided the stone for Lincoln Cathedral, while Barnack's wares were widely dispersed in parish churches, Cambridge colleges, the cathedrals of Ely and Peterborough, and the monastic houses of Crowland, Thorney, Ramsey and Sawtry.

Some other quarries did not come into production until the centuries of the medieval stone building boom. They include the Headington quarries

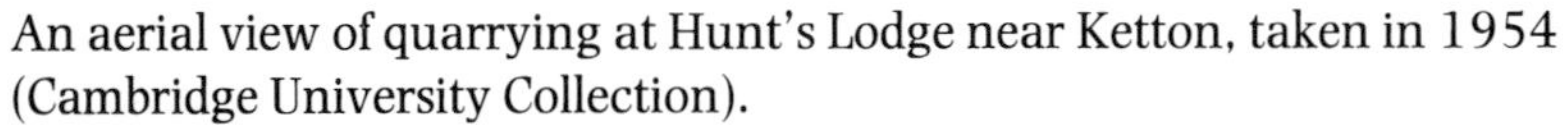

An aerial view of quarrying at Hunt's Lodge near Ketton, taken in 1954 (Cambridge University Collection).

The closure of a quarry can create its own landscape of desertion. This is part of the abandoned granite-quarrying village of Porth-y-Nant in North Wales.

in Oxfordshire, which furnished the stone for so many wonderful church and college buildings, the Weldon quarries of Northamptonshire, supplying the fine-textured oolite which was possibly used in the building of Old St Paul's Cathedral and which survives to be admired in Cambridge colleges and several noble mansions, and the Ketton workings in the same county. The Ketton stones, which clearly display the oolitic cod's roe-like texture, may have been worked in Roman times, but did not come into prominence until the closing stages of the Middle Ages. It was Barnack that supplied the stone for Ketton church, and Ketton's rise may have been stimulated by the exhaustion of the Barnack rag. In due course, the Ketton product found a ready market in Cambridge, and it is displayed in parts of many college buildings, including Clare, Pembroke, Emmanuel, Trinity, Caius and Peterhouse.

Other quarries still did not come into production until well after the close of the medieval period. They must include almost all of those in Cornwall and Aberdeenshire, which exploited the tough, prestigious granites, and the vast recent granite and basalt roadstone quarries. The Welsh slate workings only began to disperse their wares across the English lowlands after the canal and railway revolutions had provided the means for cheap transport to the swelling industrial centres.

As well as varying greatly in age, quarries naturally range greatly in size – from the 465-foot deep crater of Rubislaw granite quarry on the outskirts

An abandoned slate quarry near Keswick.

of Aberdeen, to the innumerable village stone pits of areas which are underlain by reasonable but unexceptional building stones. While the limestone belts are punctuated by famous quarrying villages, like Corfe, Taynton, Doulton, Weldon, Ancaster and Huddleston, the neighbouring village communities would often open a local quarry rather than import a costly famous product. In a stone-rich county like Northamptonshire, almost every village seems to have the scar tissues of small-scale quarrying nearby, though the workings may cover no more than the area of a tennis court, and survive only as grass-grown ripples of bumps and mounds.

Although larger and much-renowned quarries, like those of the Isle of Purbeck, Headington or Barnack, exported their product over many miles, the top-ranking quarries never obtained a monopoly in the supply of building stones. Each was surrounded by a proliferation of lesser workings, and while the master masons in charge of the most richly-funded operations might obtain prestigious stones at great expense, the choice of stone at many sites will have been determined by transport costs. Some examples have already been given, and others have been gleaned from old accounts by L. F. Salzman. In general, it seems that the cost of transporting a shipment of stone over a distance of just a dozen miles was equal to the cost of quarrying the stone. In the case of the repairs to Tutbury Castle in Staffordshire in 1314, the stones were hewn at Winshill quarry at a cost of four shillings the hundred, while the transport for the six-mile haul to the castle cost seven shillings and sixpence the hundred. In an admittedly extreme case – in the construction of Norwich Cathedral in 1287 – Salzman shows that stone was purchased at Caen for £1 6*s* 8*d*; '. . . its freight by ship to Yarmouth was £2 10*s* 8*d*; unloading it there into barges cost 2*s* 2*d*; carriage in 6 barges to Norwich was 7*s* 2d; and from the wharf to the cathedral yard was another 2*s*'. Clearly there was much to be said for working a nearby source of stone, perhaps foregoing a measure of quality, but saving greatly in costs as a result.

Different types of stone present the quarryman with different advantages and problems. Let us consider first the qualities of metamorphic rocks – so called because they have been subjected to such intense episodes of subterranean heating and pressure that their appearance and structure has been transformed. Gneiss is such a rock, often a metamorphosed granite in which the quartz, felspar and mica constituents have re-crystallized in wavy bandings. The name, which is pronounced 'nice', derives from an old German word meaning 'sparkling'. Schist, named from a Greek word meaning 'split', often derives from a sandstone which has experienced extreme compression. The mica contents of the rock may be seen to have re-formed in glittering sheets. Both gneiss, with its zebra-like bandings, and the paler types of schist form attractive rocks when they are exposed in sea-cliffs or found as boulders or amongst the shingle on a beach. Boulders of a finely-striped white schist were gathered by the makers of the remarkable prehistoric passage grave cemetery at Carrowmore near Sligo to form the brilliant circular kerbs surrounding the burial chambers. However, both gneiss and schist are unattractive to the quarryman and mason because of their toughness and uneven texture. Quartzite is another metamorphic rock which had more appeal to the prehistoric users of stone: quartz rocks form the stone circle at Duloe in Cornwall, while many monuments had a garnish of imported quartzite chips. It results from the intense heating of a sandstone so that the individual sand grains melt and re-crystallize, while the disappearance of the bedding structure results in a rock which is hard and may split in any direction. Quartzite was occasionally exploited by the makers of stone tools. Marble, which derives from the re-crystallization of the calcite in a limestone rock, is infinitely more attractive to the creators of important buildings, but occurrences of high-grade marble are rare in Britain. From the Roman era onwards, excellent decorative substitutes have been found in the form of the false marbles, most notably those of the Isle of Purbeck.

Of all the metamorphic rocks, the most extensively exploited is slate, which results from the intense compression of a shale. Apart from its hardness, the great attraction of slate is the ease with which it can be split into sheets which, being slender and therefore light, form an ideal roofing material. However, the obvious layered appearance of slate does not represent the original bedding planes upon which the layers of silt were deposited. Instead,

Proximity to a notable quarry sometimes gave rise to local craft industries and in Cornwall a local tradition of exquisite high relief carving in slate developed in some villages close to the quarries. This is part of a memorial in St Tudy church, north of Bodmin.

the 'slatey cleavage' has formed at right angles to the directions of the pressure which metamorphosed the rock. A number of slate quarries producing dark blue or grey slates line Scotland's Great Glen Fault, which runs across the Highlands from south-west to north-east and contains Loch Ness, Loch Lochy and Loch Linnhe; excellent greenish slates are worked in Westmorland, while Cornwall boasts the vast Delabole quarry, yielding tough grey, green and reddish products. The greatest concentration of large slate quarries is in Wales, and innumerable acres of darkly-roofed industrial terraces in England and South Wales bear witness to the remarkable output of the Welsh quarries in the nineteenth and early twentieth centuries. The most important workings included the Penrhyn quarries, in which the blue, grey, red and mottled slates have been extracted from benches, so that the vast resultant mountain bowl has the terraced form of oriental paddy fields. Others of note were the Dinorwic quarries, which also yielded slates of many

An aerial view of slate quarrying at Dinorwic in Snowdonia, taken in 1949 (Cambridge University Collection).

different hues; the Bangor workings, producing blue and purplish slates; those of Ffestiniog, which provided blue-grey wares; and the Prescelly quarries in the south which produced a tough grey or greenish product. Like so many other Welsh industries, slate mining has experienced a decline, and in most modern buildings slate has been superseded by composite roofing tiles. In the Lake District, the old farmsteads and some new buildings employ varieties of Skiddaw Slate in their rough and craggy walls.

While the heyday of the Welsh slate industry was in the nineteenth century and the early years of the twentieth century, when more then 10,000 men found work in the quarries, the exploitation of slate has a long history. Slate was used for roofs and floors in the great Roman military base at *Segontium*, overlooking modern Caernarfon, and slate was also used in the better qualities of northern Welsh housing in the medieval period. An ode written to the Dean of Bangor by one Guto'r Glyn in the fifteenth century mentions '. . . the stone of Gwynedd, a beautiful rock jewel from the hillsides . . .' and continues: 'We shall have slates, worn slabs, as a crust on the timber of my house.' By the early eighteenth century, an export industry had developed from the north-western seaports, while the acceleration of the Industrial Revolution created a need for mass housing in the black new boom towns of England.

The potential of the slate industry was recognised by the thrusting entrepreneur Richard Pennant, who acquired the vast Snowdonian Penrhyn estate by marriage and purchase. As the first Lord Penrhyn, he controlled an empire of quarries served by new roads, jetties and quarrying villages. The most imposing monument to the industry is Penrhyn Castle, a towering mock-Norman structure financed from the profits of Jamaican sugar plantations and Welsh slate. At the end of the nineteenth century, production peaked at almost 500,000 tons of finished slate per annum. Subsequently, changes in building methods, with the adoption of synthetic materials, have eaten away at the market for Welsh slate, and employment has slumped to one tenth of the former level. However, many Welsh slate roofs have a life expectancy of around a century and those built at the height of the quarrying boom may be at the end of their serviceable lives. Housing improvements have brought a slight recovery to the industry, although there has been severe competition from Spanish slates, imported at around two-thirds of the cost of the Welsh product. Meanwhile, a number of working

Penrhyn Castle, an extravagant mock-Norman mansion built from the profits of Welsh slate quarrying and Jamaican sugar plantations – now a National Trust property.

At the Gloddfa Ganol quarry, a terraced row of quarry workers' cottages now serves as a museum and a series of interiors have been carefully reconstructed.

slate mines – like the most visitworthy Gloddfa Ganol overlooking Blaenau Ffestiniog – have tapped the considerable public interest in industrial archaeology and gained a second income.

Being both tough and fine-grained, basalt is widely quarried as a source of road-making materials. Some of the most striking quarrying landscapes are to be seen around the basalt layers which form the summit plateaus of the Clee Hills in Shropshire. Large-scale quarrying is actively pursued beside Cleehill village, where the landscape has been gouged to produce scenery which is reminiscent of the Grand Canyon in miniature. A mile or so to the north, the basalt quarries have been abandoned and the quarrying village of Bedlam, which consists of a pair of red brick industrial terraces, lies in lofty isolation, still peopled, but by-passed by the industry which gave it birth.

Let us look next at the properties of the 'igneous' rocks, which result directly from the cooling of 'magma' or molten rock, either deep down beneath the surface of the earth – as in the case of the granite family –

or at or near the surface of the earth – as with the basalts and volcanic rocks. As we have seen, the prehistoric makers of stone axes in the British Isles were attracted by some of the outcrops of tough volcanic rocks; while beyond Britain, black volcanic glass, or 'obsidian', which has flint-like qualities, was widely used. Though huts and simple buildings were built of the granite moorstone, little use was made of granite in the construction of the more sophisticated buildings before the fifteenth century, and this intractable rock was not quarried, but obtained from moorstone blocks lying in the surface soil layers. With the improvement in quarrying and rock-cutting techniques, granite became a popular choice for the construction of prestigious buildings, many of which display the products of the quarries of North East Scotland. The wear-resistant rock was also used for more prosaic purposes: quarries like Porth-y-Nant on the Lleyn peninsula supplied paving stones to the Merseyside industrial towns.

The third great class of rocks, the sedimentaries, includes those which are composed of the eroded detritus of older rocks – the sandstones, shales, conglomerates and breccias – and others which consist of the accumulated remains of the calcareous shells of sea creatures – like the limestones and chalks. Different types of sandstone are widely displayed in the buildings and monuments of Britain; the finest grades almost rival the limestones, and the remarkable carved high crosses of Ireland are executed in fine-textured sandstones.

One of the most widely-used sandstones is the Millstone Grit of the Pennines, a tough, gritty rock which was used, as its name suggests, by the makers of millstones. However, while stones were quarried from the grits of the Peak District in Derbyshire and used for milling grist, the old millers preferred to grind fine flour using French burr stones, consisting of several stone sections cemented together and bound with an iron band. Millstone Grit is displayed throughout the gritstone districts of the Pennines in sturdy field walls and rugged farmsteads, and it was also employed in the construction of more important buildings, such as Kirkstall Abbey. It is now rather expensive, and many industrial buildings in the Dales exploited the softer sandstones of the Coal Measures series, while some new buildings re-use stone from demolished mills and industrial terraces.

Some sandstones have a russet hue, which may range from a pinkish grey through to a deep blood red, and this colouration normally indicates a desert origin. British rocks which formed under desert conditions largely belong either to the geological period known ad the 'Devonian', dating from around 350 to 400 million years ago, or to the 'Triassic' period, which prevailed between around 200 to 225 million years ago. The desert sandstones vary considerably in their texture and value to man, some being quite fine-grained and easy to work, while others, like the conglomerates which

Sandstones of different types can be seen in thousands of buildings of many ages: Red sandstone elegantly carved in the entrance to the medieval cathedral at Kirkwall on Orkney (above). Greensand, or 'carstone', which resembles burnt gingerbread, employed in the parish church at Denver in Norfolk (opposite above); note how a superior stone has been imported for the quoins, buttresses and window tracery. Red sandstone cottages at Appleby-in-Westmorland (opposite below), an interesting contrast to the slate buildings of the areas to the west and the gritstone of the Pennines to the east.

consist of large rounded pebbles set in a finer matrix, are of little value. The better red sandstones have sometimes been exploited in areas which are devoid of superior materials: Triassic sandstone, for example, is displayed in the cathedrals of Carlisle, Chester, Lichfield and Worcester, and the castle which dominates Dunster in Somerset. Exeter lies on the western margins of a red sandstone belt, but here the decision was taken to use the local sandstone as a rubble filling for the walls. The authorities owned a chalk clunch quarry at Beer near Seaton in Devon, which yielded a white stone for facing and vaulting, while the tougher grey Purbeck stone from Dorset was used in the construction of the load-bearing piers. At Beer, the cavern networks survive and contrast with the more normal open-face forms of quarrying.

Some of the red sandstones are amenable to carving and some are strongly coloured. The vivid combination of carving and a rich red hue, as displayed, for example, in the entrance to the twelfth-century cathedral at Kirkwall in Orkney, can be very striking. The ruined seventeenth-century Earl's Palace which lies adjacent to the cathedral, however, reveals the great weakness of most red sandstones: their susceptibility to erosion. This problem is also apparent in the stonework of the West Midlands cathedrals mentioned above. Nevertheless, the Triassic sandstones of the English Midlands were quite frequently exploited by small quarries, which often originated in local church-building operations. The parish church at Laxton in Nottinghamshire, for example, is built of the local desert sandstone, which is of the common type known as Keuper Marl. Attractive red sandstone cottages line the old market place below the castle at Appleby-in-Westmorland.

The high costs of transporting building stones ensured that a variety of sedimentary rocks – which might be both easily quarried and visually attractive but were distinctly second-rate as building materials – were often employed in the construction of parish churches. A bridge between the sandstone and limestone groups of rocks is provided by the Blue Lias. Its name means 'layers' in the dialect of quarrymen, and it often consists of alternating layers of fine-grained, dark blue shale and harder bands of a bluish limestone. The Blue Lias forms a belt which runs diagonally across the English Midlands, sandwiched geographically and temporally between the older Keuper Marl and the younger oolitic limestone. It forms sea-cliffs at Lyme Regis in Dorset, where the dangerously unstable rocks are famous for the fossils of ammonites and great sea reptiles which they contain, while the less famous Blue Lias cliffs at Kilve on the north Somerset coast are also rich in ammonite fossils. Although the Lias is most widely known for its fossils, it was also sometimes quarried for local church-building projects. Somerset is noted for its remarkable Perpendicular church towers, and few surpass that of North Petherton village church, which has an unusual lilac-colour

Ammonite fossils exposed in the Blue Lias cliffs at Kilve in Somerset.

resulting from the combination of red and blue tinted Lias rubble.

Another sandstone which has been quarried to yield building stones with more visual appeal than true quality is the Greensand, which forms a narrow belt lying to the north and west of the great chalk belt of East Anglia and the southern Midlands, and also forms a part of the Weald, lying within the chalky arms of the North and South Downs. Its commonly greenish hue derives from the presence of the green mineral 'glauconite', which is used as a colouring in the paint industry. Greensand, however, comes in a variety of tints, none more striking than the biscuity chocolate of the iron-rich 'Carstone'. The Greensand belt is in many places sandwiched between chalk – which may yield only flints to the builder – and strips of clay country, so that the makers of churches and cottages have often been pleased to exploit the more durable qualities of a local Greensand outcrop. The belt is a narrow one, and so the traveller may be surprised to find himself all at once in a stone-built village which seems isolated in a surrounding landscape of flint and timber-framing. This effect is most pronounced in the west of Norfolk, where the brown Carstone makes its sudden appearances. The stone can be admired in the 'burnt gingerbred' church at Denver.

One stone which is associated with the Greensand series and which has been highly valued by masons is the tough, shelly Kentish Rag. Quarried at Maidstone, Reigate and the celebrated Merstham quarries, the rag was exported quite widely to form strong lintels and surrounds for windows and doorways. As well as being used in the building of cottages and some fine, tall-towered Kentish churches, the tough and intractable rock was also used in the building of Eton College, Windsor Castle and Westminster Palace, though being difficult to dress it was most commonly employed as a wall-filling.

The very finest building stones, as well as some of the poorest, belong to the large and varied limestone family. However, both limestones and sandstones have been employed in the roofing of buildings large and small and, having already encountered the true, metamorphic slates, we may now conveniently consider the various other forms of roofing stones together. Wherever such stones have been employed, the delightful tones and textures of the roofs contribute enormously to the joy of the landscape. The products of the small 'slate' quarries of the Yorkshire Dales have already been noted. Mention should also be made of the Rough Rim flags which crown some of the older dwellings in the Halifax and Huddersfield area, and the Elland flags, Rochdale flags, Alton rock and Winfield flags of west Yorkshire, east Lancashire, north Staffordshire and Derbyshire respectively. Further afield, similar Carboniferous flags roof many tiny cottages in the southern part of County Clare. These Liscannor flags are still worked and, being patterned by fossil worm tracks, they are sometimes used in ornamental work. Although the interbedded flags and gritstones here form the Cliffs of Moher, the highest in Ireland, this geological climax is no more than an aperitif for the limestone splendours of the Burren district lying immediately to the north. Returning to England, the much younger Wealden beds in places yield sandstone flags, which are widely used in the vernacular dwellings of the south-east and are known as 'Horsham Stone'.

Limestone 'slates' were obtained from a number of small quarries punctuating the great Jurassic belt. The two most celebrated quarries were Collyweston in Northamptonshire, which provided the slates to roof many fine old buildings in Stamford and the east Midland area, and Stonefield in Oxfordshire, the source of many Oxford college roofs. The Brockhill quarry in the Cotswolds was another source, exporting its wares as far afield as Christchurch, Oxford and Carisbrooke Castle on the Isle of Wight.

The production of stone tiles in the Cotswolds ended almost three decades ago in the face of competition from reconstituted materials. As a result, the house-proud owners of Cotswolds cottages have faced increasing difficulties in finding authentic replacements. In 1981, however, an unexpected salvation arrived with the discovery of a very large supply of slates abandoned twenty-five years ago and lying just below the surface of a Gloucestershire field. Two young entrepreneurs based in Temple Guiting are spearheading a project to split, dress and market the tiles; success may result in the provision of a supply of tiles sufficient to satisfy all foreseeable demands, and in the creation of twenty new jobs in an industry which was thought to be extinct.

Limestone comes in many different forms and qualities. The oldest extensive outcrops, belonging to the Carboniferous period of around 280 to 350 million years ago, will be familiar to all visitors to the Mendips, the Derbyshire Peak District, the Pennines and the Irish coastal area between County Clare and County Donegal. The Carboniferous limestone, which tends to yield tough, angular chunks of rock, does not lend itself to shaping and dressing, and it is generally obvious in field walls, and in the whitish or bluish rough-textured walls of barns and farmsteads.

From the point of view of the mason, the aesthete and the homeowner, one cannot come closer to perfection than the oolites and ragstones of the 200-million-year-old Jurassic limestone belt which sweeps across England from the North York Moors to the Dorset coast. Many of the names of the great limestone quarrying centres – like Ancaster, Barnack, Chilmark, Headington, Ketton, Portland, Taynton and Weldon – have already been mentioned. A splendid array of medieval cathedrals, including those at Ely, Exeter, Lincoln, Peterborough, Salisbury and Wells, display their fine and varied products.

Limestones are formed from the calcium-rich shells and skeletons of many forms of marine life, and most of the British limestones accumulated under the shallow waters of tropical seas. Calcium carbonate easily passes into and out of solution, and consequently, while some limestones consist entirely of recognizable fossils, others contain few and consist of carbonate sediments precipitated from the seawater after the shells and skeletons have dissolved. The most obviously fossiliferous limestones include the crinoidal forms. These can be seen in sea-cliffs at places like Lyme Regis and Whitby, and consist of the remains of animals known as 'crinoids', or sea lilies, and in the coral limestones which are evident, for example, in Shropshire's Wenlock Edge, a fossilized coral reef some thirty miles in length. Visitors who make the rewarding ascent to the prehistoric tomb cairn which crowns the summit of Knocknarea near Sligo will not fail to notice that the surrounding field walls are composed of densely-packed and tube-like crinoid fossils, resembling grey fag-ends in an overflowing ashtray.

The Jurassic limestone belt contains a variety of ragstones with a gritty texture – which results from the concentration of tiny shell fragments – limestones which are stained gold or amber by impure ores of iron, and also the supreme oolitic limestone. This is an intensely fossiliferous limestone, although the composition of densely-packed spherical seashells known as 'ooids', each scarcely more than one millimeter in diameter, is only apparent on close inspection. To see oolitic limestone in the process of forming from the shells of ooids today, one must go to the Florida coast, where the living ooids in their countless millions are rolled and tumbled in the shallow, sub-tropical coastal waters. Medieval masons, who knew

Alabaster became extremely popular as a material for exquisitely-carved memorials in the latter part of the Middle Ages. This is a detail from the Marmion tomb in West Tanfield, Yorkshire.

nothing of either Florida or ooids, were well aware of the remarkable qualities of this easily-worked, fine-textured and durable limestone, and appreciated that the best quarries yielded a stone which was readily sawn, but which actually hardened after exposure in the walls of a building.

A special mention must be made of the exceptional but varied limestones of Dorset. Purbeck marble is a red or green coloured stone which was highly valued in the medieval period for its decorative quality, being polished and generally darkened with varnish or oil to a glossy, near-black hue. The striking patterning of this false marble derives from the closely-packed shells of fossilized freshwater pond snails. The heyday of the Purbeck marble quarries coincided with the great cathedral-building era. Belgian marble was imported to England in the twelfth century, but it was displaced by the Purbeck product to the extent that almost every important church built during the High Middle Ages contained polished piers, an imposing tomb or a font of Purbeck marble. In the course of the fifteenth century, Purbeck marble tended to be superseded by soft, white, translucent alabaster in the carving on internal church monuments and statuary, the alabaster quarries at Chellaston in Derbyshire being the most important source of supply. The Purbeck marble was quarried from two narrow seams, frequently less than one foot in thickness; medieval quarries like those near Dunshay, Afflington and Blashenwell follow the seams from Swanage to Church Knowle.

The Isle of Purbeck also yielded another valuable limestone, the Purbeck freestone, which may first have been exploited by the Romans in the construction of the town wall of Dorchester (*Durnovaria*). The Purbeck beds yielded building stones, which are displayed in the walls of local churches, and roofing slates. Near Worth Matravers, quarries were cut directly into the sea-cliffs so that their product could be loaded directly into waiting ships. The Portland limestone of the nearby Isle of Portland was of an even higher quality. Early in the Middle Ages, it was used locally, as in the construction of Rufus Castle; later it was exported for use in Westminster Palace and the Tower of London. The most important developments occurred in the post-medieval period, after Inigo Jones had selected Portland stone for the Banqueting Hall at Westminster and the York Gate at Greenwich Hospital. Wren, in turn, used the stone for the churches built in the aftermath of the Fire of London, and for the new St Paul's Cathedral. The demand for Portland stone in London was considerable, and until the development of railway connections in the dying days of the stone building era, the stone reached London by ship, often from quarries hewn into the sea-cliffs.

The legacy of centuries of quarrying in the Dorset landscape consists of much more than the scores of abandoned and overgrown quarry workings. It includes, for instance, the remains of the quays and wharves at isolated hamlets like Redcliffe and Slepe, which once bustled with the loading of stone. As Christopher Taylor points out in his guide to the Dorset landscape, the masons and the sculptors who worked the Purbeck marble were concentrated in the little town of Corfe Castle. The stone was brought here by cart and wagon from the numerous medieval quarries, and then the shaped stones and marble sculptures were conveyed overland to the southern shores of Poole Harbour for export. Several of the lanes which were used survive to this day, deeply engraved in the landscape by the creaking wheels of the laden stone wagons.

Many of the advantages of the Jurassic oolites are shared by the older, whitish beds of the Magnesian Limestone, which form a slender geological belt of bleached limestone running southwards from the Tyne estuary and terminating about five miles to the west of Nottingham. Travellers moving eastwards from the Millstone Grit country through Knaresborough and towards York will notice the paleness of the stones in the villages which line the western margins of the Vale of York, contrasting with the darker tones of the gritstone villages of the Dales. The Magnesian Limestone sometimes has a yellow or buff tint, which derives from the presence of dolomite, a carbonate of magnesium and calcium. Of the numerous quarries which worked this amenable but durable rock, the most notable was at Huddleston

The castle and quarrying village of Corfe in Dorset.

near York. It supplied many of the stones used in the construction of York Minster, but also dispersed its wares more widely, for use in Eton College and King's College Chapel, Cambridge. Magnesian Limestone can also be seen in the Houses of Parliament and Beverley Minster.

The youngest, softest and probably the least prestigious of the British limestones are the chalks, which date from the Cretaceous period of around 100 million years ago. The chalk beds, almost half a mile thick in places, consist of minute particles, only around three hundredths of a millimeter in diameter, which represent the plate-like armour of a tiny, spherical form of algae known as '*coccolithoporid*'. The chalk deposits contain horizons rich in flints, which probably derive from the skeletal debris of sponges, while the chalk beds themselves are not of a uniform softness – harder beds can often be seen jutting forth as minor escarpments within the chalk itself. While flints gathered from field scatters were incorporated in Roman town walls, their forts of the Saxon shore and a large number of Saxon and medieval churches, the chalklands are by definition poor in stone, and masons were pleased to exploit the harder bands of rock within the chalk, which yield the white building stone known as clunch.

Although less robust than other limestones, clunch is very fine-grained and easy to work and carve, and may be quite acceptable for internal church architecture, as displayed in Norwich Cathedral and the Lady Chapel at Ely. In a number of lesser parish churches, like the one at Landbeach in Cambridgeshire, clunch was used for the piers which perform the stern task of supporting the roof. Even in this role, the rock is sufficiently soft to be scratched with ease and the grafitti contained in a clunch-lined church can be fascinating. (Recent additions are known as 'vandalism', older scratchings as 'historical evidence'.) At Landbeach, for example, one of the piers carries what appears to be the groundplan for a medieval aisled hall and, with a little imagination, we can visualize the prospective owner and the housewright lingering after mass to discuss the placings of the timbers. Clunch will not withstand the effects of friction and it weathers quite rapidly when exposed in the outer walls of buildings. Even so, it provides a number of dazzling white chalkland church towers which endure in medieval clunch-quarrying villages like Barrington and Cherry Hinton, both in Cambridgeshire. Hundreds of settlements in the chalk belts were served by their local clunch pit, that which furnished the needs of Great Shelford near Cambridge lying completely forgotten quite close to my former cottage. A fine example lies behind the clunch-built church at Orwell in Cambridgeshire. Of the larger workings which produced a good-quality clunch, the most

The medieval clunch quarries at Aston Clinton in Buckinghamshire now serve, like several other abandoned chalk workings, as a nature reserve noted for its wild orchids, although the hummocky terrain still reveals the former function.

noted was probably that at Totternhoe in Bedfordshire, which provided some of the stones used in the building and extension of Windsor Castle.

Finally, of the various imported stones which were luxurious by the very virtue of their importation, the most significant by far is the creamy buff oolite of Caen in France. Like the best of the English oolites, the Caen stone hardens after exposure to the elements, and is also amenable to quite intricate carving. It was imported shortly before or just after the Norman Conquest, when several of the best native stones were still untapped. Caen stone was possibly partly used in the building of old St Paul's and St Alban's Abbey, and also in the rebuilding of Canterbury Cathedral; consignments continued to arrive in England throughout most of the Middle Ages.

Chapter 9

In the Quarry

Given the rich variety of the stones which were employed in different places at different times by British masons, it is not surprising that quarrying techniques were also varied; I will therefore attempt only to highlight specific examples of contrasting quarrying methods.

Apart from the various legal problems of tenure which might occasionally restrict operations, the most important technical restriction which affected medieval quarrying ventures was the lack of an effective means of removing water from the deeper workings. This problem was probably far more burdensome than the medieval quarryman's lack of dynamite. Although short, vertical pits or shafts and longer horizontal adits were quite frequently employed in the extraction of stone, the most favourable conditions for quarrying existed when the desired rock outcropped in the form of a cliff-like scarp face. As continued working caused the open quarry face to be eaten away, a horseshoe-shaped scar would develop in the scarp, with shallow ramps of rubble leading back from the foot of the workings towards the lanes plied by the stone carts and wagons. However, while such classically simple fossilized quarrying landscapes may sometimes be formed, in general the old quarrying landscapes tend to be more chaotic and confused. Where, as at Barnack, quarrying has been practised over a number of centuries, the pits and spoilheaps of abandoned workings may be inextricably entangled in a petrified storm sea of chasms and mounds; in many other places, serious or half-hearted efforts to backfill or reclaim the pit and spoil-scarred landscape may result in an unintelligible pattern of troughs and swells.

The quality of rocks can vary quite considerably within a small area, and the determined pursuit of a narrow band of a particularly desirable layer may result in a landscape patterned by a string of pockmarks which trace the course of the seam. The qualities of a stone vary even within a single quarry. Perhaps it was the proximity of the Headington quarries to Oxford

Basalt, a tough igneous rock, is sometimes quarried as a source of roadstone. This is one of the Clee Hill quarries near Ludlow.

which encouraged the use of the poorer as well as the better Headington stones in the various college buildings – with unfortunate consequences both for their fabric and for the restoration accounts. On the whole, however, the medieval masons were well aware of the qualities of the different stones exposed in a quarry, and they would make it their business to visit the site to inspect the better stones and discard the poorer. In the case of the larger undertakings, it was common for masons to be permanently present in the quarry, selecting the good stones and cutting, dressing or 'scappling' those which they had chosen.

When, as was commonly the case, the desired stone was not exposed in a scarp face, the first stages of quarrying involved stripping away the overlying blankets of soil and sub-soil and then removing the surface layers of rock, which were likely to be fissured by roots and frosts and rotted by humic acid seeping downwards from the soil. Although the tasks allotted to masons and quarrymen frequently overlapped, such work was normally accomplished by low-paid, unskilled labour. The next stage involved the extraction of large blocks of stone, which were split from the bedrock. In cases where the bedding planes were near-horizontal and well-developed, the blocks could be levered from the strata after vertical cuts had been made using lines of wedges which were hammered home with heavy malls until a fracture resulted. Then, depending upon the nature and hardness of the rock, a further reduction of the severed block into building stones would be achieved by more splitting and scappling with a heavy hammer, or by sawing. In some cases, the final stages of dressing stones, and even shaping components according to templates, was accomplished in the quarry, just as the housewright would prefabricate the components of a timber-framed building in his workshop. In consequence, no young labourer could work for long in one of the larger quarries without gaining an awareness of the craft of the mason; as a result, the medieval quarries were the nurseries and training grounds of the building industry.

Since several, and sometimes all, of the stages in the cutting, dressing and shaping of stones were performed in the quarry, the tool-kits of the mason and the quarryman tended to overlap, and it is convenient to consider all the stone-working tools together. In the pre-industrial era, the tools and devices employed were relatively simple, and although those used at different types of quarry might differ in detail, there were many basic similarities. In quarrying the softest stones, the wedge technique described above would normally be employed, with small iron wedges, or 'gads', being driven home to cause vertical splits around the margins of a chosen block of stone. Next, gads were driven along a bedding plane to create a horizontal split severing the block, which would then be levered from its setting with crowbars. The block might be removed by means of ropes and levers or simple cranes to the part of the quarry where it could be cut and shaped. In due course, 'nippers', resembling a gigantic pair of forceps with inwardly-curving points, were employed in the lifting of the stones. As the ropes or chains from the crane arm pulled on the ends of the nippers, so the points bit deeply into the stone. The heavier the block, the tighter their grip. When the heavier rocks, like granite, were being quarried, a still more powerful hold was needed, and this could be achieved with 'lewises'. A widening hole was cut in the block and the fan-shaped metal devices were inserted; as the crane hauled on the lewis, so the lewis spread to fill and grip the sides of the cavity, like the components in a dove-tail joint. The tougher rocks were often split from their setting using 'wedges and feathers', rather than simple wedges. A series of holes was cut along the chosen line and a pair of feathers, each member shaped like a pointed shoe horn, was inserted in each hole. A wedge was then driven between each pair of feathers to ensure the splitting of the stone, rather than the jamming of the wedge.

Once removed from their setting, the clunches and soft oolites could be sawn into shape using hand saws quite similar to those used by carpenters. Larger blocks might be cut using a frame saw, which was kept in a true position by a simple set of pulleys; sharp sand and water ran in the saw groove to assist the cutting action of the heavy steel blade. Smooth-faced ashlar blocks could be sawn directly from the softer building stones without needing any shaping with hammers. Tougher rocks were rough-shaped, using the square-faced 'spalling hammer' or the 'scappling hammer', which had one pointed face. During the first century of Norman rule, the axe was used for the shaping and even the carving of building stones, but during the twelfth century, it was increasingly superseded by the broad chisel known as a 'boaster' or 'bolster', which was driven across the face of the stone by blows from a rounded wooden mallet. In the course of that century, the flat-ended boaster also tended to be replaced in rougher work by the 'claw tool', which had a serrated edge. The use of different shaping techniques can often be recognized in the tooling on the surface of a stone. Where the scappling hammer alone has been used, then the surface has a rough, cratered appearance; the axe leaves a smoother but still irregular surface with uneven scars; the boaster tends to leave a series of parallel grooves which chart the movements of the tool across the stone face; and when the claw tool was used for the final shaping of stones, a slightly pitted or spotty appearance results. The later medieval masons tended to tool stones diagonally rather than vertically across their faces. For finer work in the carving of mouldings and for decorative work, the masons used a 'dummy' – a smaller version of the beechwood mall – and metal chisels with narrow points of various shapes, while wooden-handled chisels were used on the softer stones.

Finally, there were the setting-out tools: the 90° square; the triangular set square; the 'bevel' of two adjustable metal blades, which was used for measuring angles; compasses; and the plumb rule and line, used to test the straightness of work and to achieve the vertical. A simple water table device was used to test the true horizontal, for the spirit level was not invented until the seventeenth century. Thus, with tools as simple as these and the various sets of standardized templates for testing the profiles of mouldings, medieval craftsmen were able to construct stone buildings ranging from the humble church to the most magnificent cathedral. As for plans, these might be no more than visions in the minds of the master masons and their employers, or mere sketches scratched upon softwood planks, which could be carried around the building site in all weathers.

Quarries may be fascinating, but many lie in inaccessible localities and so can be distinctly dangerous places. In recent years, some of the most interesting and dramatic quarries have been opened to the public – the majority of these were concerned with the extraction of true slates as roofing materials. The landscapes of slate quarrying are varied, their characteristics being determined by the scale of the operations and the nature of the local geology. Normally, slate was quarried by open-cast methods, and in the

A demonstration of slate splitting and shaping at the Delabole quarry in Cornwall.

largest quarries great amphitheatres resulted, with spiralling roads following the stepped benches cut into their walls. In other cases, however, the seams of top-grade material were exploited by subterranean tunnels. The Carnglaze quarries near the village of St Neot in Cornwall are now open to the public. Here, the extraction of the high-grade blue roofing slates began with open workings cut into the steep flanks of the River Loveny valley. However, the need to extract only the best quality roofing material resulted in the accumulation of great mounds of discarded dross. Dumping grounds for waste material were limited by the topography, and it became apparent that quarrying by underground tunnel techniques would remove the need to dispose of great quantities of useless surface overburden and give access to the deeper seams of top-grade material. Fortunately, the history of copper and tin mining in the locality had provided a reservoir of skills in tunnelling techniques, while the stability of the Carnglaze slate allowed great underground chambers to be excavated with minimal amounts of rock being left *in situ* to provide supporting pillars.

The date of the commencement of quarrying at Carnglaze is uncertain, but it probably lies in the late eighteenth century. Slate-working in Cornwall has a much longer history than this, for by the fourteenth century, houses in the county were commonly roofed in slate, and there may have been a small-scale export trade from small Cornish ports. The magnificent Delabole workings were in production at least by the sixteenth century and, while exploitation continues, a public exhibition centre is also displayed at the Delabole. Transport difficulties, however, greatly restricted the scale of quarrying operations until the improvement in road-building and the canal and railway revolutions opened the English market to the Welsh and Cornish quarries during the eighteenth and nineteenth centuries. In the case of Carnglaze, the dressed slates were at first taken by pack-horse trains to the ports of Looe and Polperro. But the construction of the bridge at Saltash in 1859 and the arrival of the Great Western Railway allowed the slates to be carted just two miles to Doublebois Station and conveyed by rail for export; the blue Carnglaze slates also provided many of the roofs in the ports of Plymouth and Penzance.

The extraction of slate from tunnels and caverns was also the chosen method of quarrying at Llanfair near Harlech in Wales, where the old workings are now open to the public. This is a relatively young slate quarry, having been opened in 1877 at the height of the boom in demand for Welsh slate created by the transport revolutions, and by the unprecedented requirement for cheap housing that followed the Industrial Revolution. The first workings to respond to this demand on a grand scale lay mainly at places like Bethesda and Llanberis in Caernarfonshire, where the vertical orientation of the roofing slate beds favoured open workings cut into the hillsides. In Merioneth, however, the beds tended to be horizontal, creating quarrying problems similar to those encountered at Carnglaze, so that the tunnelling technique was favoured. The Llanfair quarry ceased production in 1906, but in the meantime a great network of underground passages and caverns had been created. The materials extracted had travelled not only to the youthful industrial centres of England, but also, in the holds of the bustling hordes of vessels which exported slates from Portmadoc docks, to Baltic and Canadian ports.

The initial stage in the exploitation of a Welsh quarry involved the discovery of an economic bed by an 'adventurer', and his endeavours would be followed by a quest for development capital before work in the quarry could begin. Two different groups of skilled workers operated in slate mines like those at Llanfair. Firstly, there were the miners, who opened up the workings by driving drifts, or 'roofs', into the slate beds to create a series of tunnels which were parallel to each other and which also penetrated the rock at different levels so that the wide pillars left to support the ceilings of one line of slate-working chambers lay directly beneath the supporting pillars of the higher level workings. Then, the slate-getters, or 'rockmen', entered the workings, enlarging the chambers by drilling holes in which gunpowder could be inserted to free the slate, block by block. (Mechanical drills were introduced in the mid-nineteenth century.) As the workings were widened, narrow-gauge trackways were introduced – plied by small slate wagons, which were manhandled and hauled by pulley systems along the

Blaenau Ffestiniog is ringed by slate quarries, some working and others abandoned. This is the view from the Gloddfa Ganol quarry, which is both working and open to the public.

inclines and levels to the cutting sheds that lay outside the mine. The blocks of slate arriving from the quarry were first cut to size across the grain of the cleavage, using a steam-powered rotary saw. Then, the slender roofing slates were split from the sawn blocks by a craftsman using a mallet and broad chisel. In the older quarries, these finishing processes were carried out in rows of three-sided, slate-walled sheds called 'gwalia', where the craftsmen worked at cutting and splitting. Rockmen alternated their work between spells at the rockface and spells in the gwalia. Later, the gwalia were replaced by mechanical dressing and splitting mills.

As with the 'false' slates of Collyweston, Stonefield and the Yorkshire Dales, many different sizes of roofing slates were manufactured, but the Welsh products came to be known by different names of a feminine and regal nature, with Small Ladies being the smallest slates, and the range of sizes ran up the hierarchy through Broad and Wide Ladies, Viscountesses and Countesses to Duchesses and finally Princesses, which measured twenty-four inches by fourteen, and Empresses of twenty-six by sixteen inches. As a Welsh judge wrote early in the nineteenth century:

> By a stroke of the hammer, without the King's aid,
> A lady, or countess, or duchess is made.

Other old slate workings are open to the public in the area of Blaenau Ffestiniog, which lay at the centre of a complex of highly productive nineteenth-century quarries; a number of active workings remain. Nearby are the Llechwedd slate caverns, where visitors can see sixteen different levels of tunnel and chamber workings, while the Mountain Tourist Centre conducts tours through old mine workings by Land Rover. Much of the original slate cutting machinery is preserved at the Oakeley Slate Mill, where craftsmen demonstrate slate-splitting techniques.

In order to appreciate the local and historical differences between various types of quarries, let us now take a closer look at three contrasting quarries: the Weldon quarries in Northamptonshire, those at Stonesfield in Oxfordshire, and the Aberdeen granite quarries.

The Weldon stone is a first-rate oolite which was particularly valued by the carvers of medieval church tracery and mouldings. When freshly quarried, it has a buttermilk hue, but it weathers to a silvery grey. It has all the ideal qualities of the best oolites, being softer even than Bath stone when fresh, or 'green', and easily sawn to shape. Furthermore, the prominent bedding planes enabled the least skilled layers to set the stones properly – in the positions in which they were quarried – and these, together with the numerous vertical fissures or joints, eased the extraction of stone. Of even greater value is the fact that Weldon stones, like other fine oolites such as the Caen stone, will 'case harden'. As the water content or sap evaporates from the quarried stone, a thin layer of crystalline limestone is deposited upon its surface to provide a hard, weather-resistant skin, which in modern times has been found more resistant than most limestones to the chemical decay caused by the acidic atmospheres of industrial conurbations. Recent examinations of the surviving strata at Weldon show the good oolites in beds just a few feet thick and separated by other beds of inferior building stone; and so it is likely that the medieval quarrymen were obliged to remove the less valuable, if useable, shelly ragstones in order to exploit the oolite.

The remarkably long history of quarrying at Weldon seems to have begun in the middle of the thirteenth century, although there is a local tradition that Weldon stone was even incorporated in the eleventh-century construction of Old St Paul's. Perhaps Northamptonshire folk who had seen the cathedral before its destruction in the Fire of London mistook the Caen stone for the similar Weldon product – just as the beautiful Eleanor Crosses which survive at Geddington and Hardingstone in Northamptonshire have been mistaken for Caen stone work. As Jeffrey Best, Susan Parker and Christine Prickett point out in their study of the Weldon and Ketton industries, it may well be that the oolitic freestones of Weldon were reserved for special purposes and prestigious projects, since the church at Weldon and others at

The church at the Northamptonshire quarrying village of Weldon employs stone from the neighbouring quarries at Stanion.

Geddington and Deene nearby appear to use stone from the quarries at Stanion. The reputation of the Weldon stone must have spread, and in the early 1480s it was being used in the building of King's College Chapel in Cambridge, superseding the Magnesian Limestone which had travelled all the way from the Huddleston quarries in Yorkshire. In the construction of both this chapel and Great St Mary's church, which was rising on the opposite side of the street, Weldon stone combined with that of the Clipsham quarries in Rutland. A great expansion in the use of Weldon stone took place in the sixteenth century, with orders for work on Caius College and

King's College chapel in Cambridge incorporated stone from the Huddleston quarries in Yorkshire, the Weldon quarries, and Clipsham stone from Rutland.

the building of Jesus College, while at Trinity College, the Weldon material provided dressings. (Walling materials for Trinity were robbed from the deserted fenland abbeys.)

Towards the end of the sixteenth century, the rising aspirations of the wealthier classes were finding expression in the construction of imposing country mansions, and the urge to rebuild in stone spread to the lesser gentry during the course of the following century. Weldon stone satisfied the needs of many of the hall builders. It is displayed in magnificent mansions like Kirby Hall of the 1570s and the extensions to Lamport Hall of the 1730s, as well as smaller, more homely halls like Haunt Hill House in Weldon itself of about 1640. In the early decades of the nineteenth century, when fashion ran strongly in favour of the stones of Bath and Portland, the quarries may have served only the needs of local house and cottage builders, and may periodically have been abandoned completely.

In the latter third of the century, however, they became more active and were given a great stimulus by the opening of a railway link between Kettering, which lies just eight miles to the south, and Manton in Rutland, with a connection to Nottingham. Hitherto, Weldon had suffered in competition with other quarries which were well-served by the developing rail network, but now Weldon had a station within a couple of miles of the quarries. Thus encouraged, the owner, the Earl of Winchelsea and Nottingham, imported skilled quarrymen from Bath and Corsham in Wiltshire, while a party of architects was feted at Kirby Hall.

In this way the Weldon quarries enjoyed an Indian summer which coincided with the dying days of the stone building era. In the decades which bracket 1900, Weldon stone was used for traditional purposes in work at Cambridge colleges like Sidney Sussex and Caius, and in the building of local and fenland churches as at Kettering and Ely; but it also travelled far afield, appearing in churches in Nottingham, Bradford and Dover, the central tower of Rochester Cathedral, grammar schools in Bridlington, and in the Sheffield Royal Infirmary. Production continued to rise steeply until a peak was achieved in the early 1920s, with the demand for fine freestones in post-1918 war memorials playing a considerable part.

Despite its reputation, however, the Weldon stone could not continue to find markets in a century when the tide was running against the construction of stone buildings. There were other more specific problems: the scarcity of motor transport vehicles of the early post-war period caused schedules to be broken, and a decline in the quality of stones which the two Weldon quarries were producing gave rise to complaints. It seems that much of the stone being sold was coarser and less weather-resistant than before, although this may have resulted from the loss of skilled and discerning quarrymen in the course of the war. By 1961, only two quarry workers were employed; a modest revival followed, and now reserves of excellent Weldon freestone are available for repair and restoration work at this site where quarrying has a quite exceptional 700-year pedigree.

Stonesfield village and its surrounding quarries lie ten miles to the north-west of Oxford. They were famed, not for the production of building stone, but for their roofing slates. These false slates were quarried from a material known as 'pendle', a flaggy grey sandstone which is rich in lime, fossils and ooliths. Some of the slates derived from blocks of the pendle, others from spheroidal formations which occur within the pendle and are known as 'potlids'. In common with the other famous slate industry at Collyweston, but unlike other slate industries of the Cotswolds, Stonesfield produced 'frosted' slates. Potlids and pendle were quarried and then stored, often in an old mine shaft, to prevent the evaporation of their water content or sap, which could never be recovered once it had been lost. Then, in winter, the blocks were spread upon the surrounding fields so that the stones would be split into slates by the action of frost – the internal pressures which resulted from the freezing and thawing of the sap causing them to split along their bedding planes. If the frost had done its work, the pendle blocks and potlids were said to rattle when tapped, but if the winter had been too mild to achieve an effect, then the stones had to be buried and turfed in storage for the next winter.

The origins of the Stonesfield industry are uncertain. Evidence from nearby villa remains at Ditchley and Shakenoak shows that unfrosted slates

Stonesfield in Oxfordshire was noted for the production of roofing slates. This is one of the village roofs.

from the Stonesfield locality were used in the Roman period. However, as M. A. Aston points out in his fascinating recent study of the industry, there is no clear-cut evidence of the working of Stonesfield slates in conjunction with the frosting process earlier than 1676, though by this date, the village industry seems to have become established. Writing at this time, Dr Robert Plot told of the excellent roofing material at Stonesfield, 'where it is dug first in thick cakes, about Michaelmas time, or before, to lie all the winter and receive the frosts, which make it cleave in the spring following into thinner plates, which otherwise it would not do so kindly'.

In 1705, the manor of Stonesfield was granted to the Dukes of Marlborough, along with the mineral rights of the estate; the rights to search for stone were, in turn, granted to certain individuals, and rights to work a pit were similarly leased by the Dukes. The industry provided employment for two classes of workers – the slate diggers, probably part-time workers who turned to mining during the slacker periods of the farming year, and the slate makers, who worked continuously throughout the year on shaping the slates. Although the physical evidence has now been obscured, it seems certain that the earlier diggings consisted of horizontal galleries, or adits, which were worked back into the hillsides around the village. In the latter part of the eighteenth century, vertical shafts seem to have been introduced to share the quarrying landscape with the hundred and more galleries. By the second half of the nineteenth century, it seems that mine shafts leading down to galleries had completely superseded the adits following the exhaustion of the most accessible pendle deposits.

The shaping and finishing of the frosted slates were accomplished by specialist 'slatters' or 'crappers'. A knife-like tool known as a 'zax' or 'sect', with a point on the upper side of a blade which was offset from its handle, like that of a mortar trowel, was used to free the frost-loosened slates from the blocks of pendle and potlids. Then, sitting with a stone or iron platform between his knees, the slatter used a hammer shaped like a small double-headed adze to clear the face of a slate of loose irregularities. Unlike the familiar rectangular Welsh slates, the Stonesfield slates have one rounded end and are thus shaped like a luggage label. Such rounded ends exist naturally on slates derived from potlids, and these had one end trimmed to a flat surface. Other slates which derived from pendle blocks were trimmed to a similar shape, again using the slate hammer. The final stage in the making of a roofing slate involved chipping a hole in its rounded end using a pointed 'slat pick' or 'pittaway'. In the laying of a roof, these holes held the pegs made of sheep bones or oak which secured the slates to the wooden laths of the roof.

Although the work of the slatter was relatively simple, craftsmanship was evident in the speed with which it was performed. To avoid unnecessary breakages, the slatter must have been able to judge in an instant the qualities of each individual slate. It is said that the wear on the slate picks was so intense that each one had to be sharpened by the village blacksmith every evening, at a rate of one penny per pick. The slatters normally worked in the open air, close to the quarrying sites, sitting on straw mats in the shelter of simple, tent-like booths of wickerwork or straw. The chippings left from slate-making accumulated in great piles or banks which are still prominent around the village. As with the flagstone industry of the Yorkshire Dales, roofing slates were produced in up to twenty-seven different sizes, which were measured against graduated slate sticks. The names of the different grades do not seem to have been standardized, but they included 'muffities' of nine-and-a-half inches and 'short cussems' of eight-and-a-half inches, up to 'long sixteens' of twenty-four inches. The 'all up' of six-and-a-half inches was the shortest slate produced; the slates were laid upon a roof in graduated courses, with the shortest at the ridge and the longest at the eaves.

The causes of the decline and abandonment of the Stonesfield industry are not entirely clear. At different stages in the nineteenth century, there may have been successions of mild winters, causing the failure of the frosting process, but competition from the true slate industry of Wales must have played an important part. In 1831, William Pitt removed the tax on Welsh slate and, with the development of the railways, large supplies of cheap, easily-layed and standardized true slates became widely available. Few would agree that the regular courses of blue-black or purple slates roofing some Oxford colleges are as beautiful as the less regimented oyster-shell textured brown and grey Stonesfield slates sported by neighbouring colleges. Often, we come to appreciate the value of our home-grown products only when they have been destroyed. However, sufficient Stonesfield, Collyweston and Cotswolds slates survive in the Midland counties to provide the crowning glories of numerous village landscapes – not least those of Collyweston and Stonesfield villages themselves.

Although seams of pendle remain at depth in the Stonesfield vicinity, only the Spratts Barn pit remained active at the start of the century, out of the hundred and more shafts and adits which existed a century before. By 1905, Alfred Ryman was the sole surviving practitioner of the arduous task of slate digging, and within a few years the Stonesfield industry was extinct. Welsh competition cannot have been entirely to blame. The industry had depended heavily upon the part-time efforts of agricultural workers, and the mechanization of farming along with falling farm prices will have encouraged many erstwhile workers to emigrate; and quarrying work is itself extremely rigorous and poorly paid. The Spratts Barn pit is now boarded over but is visited by villagers once each year.

The Aberdeen granite industry provides a third and contrasting example of quarrying. It is a much younger industry than those of Weldon and Stonesfield, and is associated with a much tougher and less compromising rock. Although one popular geology book presents a photograph of the late-medieval university chapel of King's College as an example of a granite building, it is in fact of sandstone. Before the eighteenth century, granite buildings will have been very rare in what became the 'Granite City'. Granite boulders were exploited in a few medieval buildings in the north-east of Scotland, such as St Machar's (fortified) Cathedral in Old Aberdeen and the later castles of Crathes, Midmar and Fraser of Aberdeenshire, but the Town House of 1721 was among the first of Aberdeen's granite buildings. In 1741, a large part of the city was devastated by fire, and much of the rebuilding and expansion of the latter part of the century was wisely carried out in granite.

A granite quarry at Rubislaw in the outskirts of the modern city had begun production by the time of the great fire, and in the course of the eighteenth century the export of granite in the form of deliberately quarried stone superseded the export of boulders gathered from field scatters and the seashore. As the Industrial Revolution gathered momentum, there was an increasing demand for granite, which could be used as a rugged constructional material in major engineering works, or polished and used as an elegant and imposing monumental stone. This latter use dates from around 1820, and is said to have been inspired by the examples of Egyptian polished granite displayed in the British Museum. Aberdeenshire granites are to be seen in engineering works like Portsmouth docks, the Bell Rock Lighthouse, and several of the Thames bridges, while the polished granites grace a host of important public buildings, including the Opera House in Covent Garden, the Stock Exchange, the Foreign Office and Australia House.

The county yielded a variety of granites and, while all consist of the large, slow-cooled crystals of quartz, feldspar and mica, their colour is quite variable. The Rubislaw granite is a silvery grey, that of Peterhead a deep pink, while granite from the Kemnay quarries is paler and more finely speckled than the Rubislaw product. The Kemnay quarries opened in 1858 and yielded good 'cube stone' for architectural work for around a century. It was only when the 480-feet deep workings at Kemnay were exhausted that Rubislaw could aspire to the title of the world's deepest quarry. The neo-Gothic extravaganza of Marischal College at the site of the city's second university flaunts Kemnay granite deep in the heartland of Rubislaw country.

The late emergence of the Aberdeenshire granite industry can be explained by difficulties associated with quarrying and working the stone. The granite is overlain by great thicknesses of glacial boulder clay, while the good quality stone lies deep within the granite mass. The quarrying of the tough rock presented many problems – and the pneumatic drill did not become available until the 1890s. But the greatest difficulties concerned the extraction of the granite blocks from the depths of the large quarries. Until Rubislaw quarry had reached a depth of 200 feet, the stone was removed by horse carts which plied a winding trackway cut into the side of the quarry. In 1873, a suspension cableway with a travelling carrier known as a 'Blondin' was developed at Kemnay and widely adopted to ease the extraction problem. Even so, without the excellent port facilities which were available at Aberdeen and Peterhead to assist the export of stone, it is doubtful if the industry could have expanded and made its wares so evident in the capital's Victorian public works and buildings.

In the course of the twentieth century, the industry evolved according to changing demands and circumstances. With its export trade and highly developed polishing and cutting facilities, Aberdeen began to import Scandinavian granites for finishing before 1900. The increase in demand for granite gravestones in black and green as well as the pink and greys of Aberdeenshire, and the subsequent fashion for using black granite in conjunction with modern architecture and materials, intensified the import trade: by the middle of the century, almost half the granite processed in the city's workshops came from overseas. At the same time, the development of quarries in Kirkcudbrightshire and the north of England, which lie closer to the main centres of demand, have eroded some of Aberdeen's traditional markets. Twentieth-century stone saws and newer wire saws, which cut at the unprecedented rate of nine inches per hour, have both reduced employment in the industry and tended to concentrate the activities in a few, well-funded operations at the expense of the smaller workshops. Meanwhile, restrictive tariffs and the weakening of old Commonwealth ties have undermined the export trade, and foreign granite suppliers have penetrated the British market.

In the late 1950s, the firm of John Fyfe developed a synthetic granite known as 'Fyfe stone', which consists of granite chips in a cement matrix and which proved to be a popular and cheaper alternative to the natural polished stone. Fyfe stone has made many appearances in the Granite City itself, and the rapid growth associated with new housing schemes, university expansion and the North Sea oil boom have also resulted in the appearance of unsightly tower blocks of concrete and brick. The great Rubislaw quarry is now abandoned, and for the past decade it has presented the city authorities with the problem of what to do with an enormous crater which is more than 460 feet in depth. The proposal that it might be used as an almost bottomless municipal refuse tip has been rejected by the owners, but the expansion of buildings towards the margins of the quarry is now adding urgency to the problem, safety being a prime concern.

Anyone who has visited a stone-built town like Richmond, Stamford or Bath will be aware of the profound contribution which the local building material makes to the appearance and ethos of the townscape. In the villages and market towns of the oolite belt, the stone endows a mellow warmth, as it also does in Oxford, where the more stately qualities of the college buildings provide style and stature too. In Kirkwall, the pinks and buffs of the local sandstone help to temper the bleakness of the leaden skies and sea frets of the Orcadian winter. In Aberdeen, however, the silvery granite buildings reinforce the mood of the elements. The city has two popular names: 'The Granite City' and 'The Silver City by the Grey North Sea' – both are apt.

My final story about the landscape of quarrying concerns the ways in which the products or by-products of quarrying may sometimes find niches in the landscape via the quarrying families.

Although they have long since disappeared, nineteenth century Stonesfield contained a large number of little shops and stalls selling the fossils which quarrymen had discovered in the course of their work. It is an attractive village today, and was probably at least as appealing in 1860, when Augustus Hare dismissed it as: 'That wretched little village in an exposed situation [which] consists of a succession of fossil shops containing specimens contained in it.'

As a result of recent researches by a retired lady, Mrs Gwenno Caffell, on her return to her native Tregarth in North Wales, a new form of Welsh folk art has been recognized and highlighted in *Current Archaeology* and the BBC Chronicle awards.

In the mid-1820s, the adoption of new machinery in the Welsh slate quarries allowed the extraction of much larger blocks of stone. Some of these slates, up to seven feet in length, were expropriated by the quarrymen and incorporated into their cottages in the forms of fireplace surrounds or, more occasionally, shelves or benches. An art of carving decorations upon the slates swiftly developed in a variety of different styles. Some involved geometrical motifs borrowed from eighteenth-century furniture or harking back to the cup-and-ring marks of antiquity. Others commemorated notable events, like the opening of the Menai Bridge in 1826, embodied intricate astronomical and astrological patterns, or recorded the music of hymn tunes. Much of the carving is of a remarkable quality and execution. The art, however, was short-lived, and all the fifty-three dated pieces discovered by Mrs Caffell date from the years 1832–43. By the 1840s, the Victorian taste for decking the fireplace with tassels and fringes from the new textile factories had penetrated the working-class concept of home, and carving ceased. Mrs Caffell's work, however, suggests that a potentially productive and rewarding hobby beckons any reader living close to a quarrying district, for each abandoned quarry had its story, its culture, its technique and its craftsmanship.

Much remains to be discovered before the dead industries are completely buried and overgrown and all memories of local quarrying are extinguished. Meanwhile, at the sites of several former Welsh slate quarries, and most notably at the former granite quarrying village of Porth-y-Nant, complete, deserted villages slumber and crumble as memorials to a once vital industry.

Much has been written about our grand old buildings from the perspective of Fine Art – a perspective which often overlooks the purpose of the building and its origins. Much less has been written about quarrying. Yet without the quarries, there would be no great cathedrals, churches or stone-wrought mansions. Abandoned stone mines are often overgrown, occasionally dangerous, and seldom places of great beauty, yet they contain a wealth of interest.

The stone from the Stanion quarries in Northamptonshire has been mistaken for Caen stone, imported from France. Here, Stanion stone is displayed in the medieval Eleanor cross at Geddington in Northamptonshire, erected by Edward I to commemorate his Queen, who died in 1290.

Chapter 10

The Early Springtime of Stone Building

With varying degrees of reluctance, the turbulent Celtic tribal kingdoms of England and Wales were taken into the Roman imperial fold in the years following the invasion of AD 43. As the legions advanced westwards, these multi-national forces will doubtless have been impressed by the crumbling monuments to a distant antiquity at Avebury and Stonehenge. It is doubtful whether the Romans will have found much in the way of contemporary English masonry to admire. In the stony uplands, they will have seen many little clusters of thatch-cone crowned drystone huts, and drystone revetments which faced the rampart banks at some of the native hillforts. They will certainly have admired the Celtic metalwork – the flamboyant neck rings, or 'tores' of twisted gold alloy wires and the weaponry, with its inlays and engravings. They may have toyed with the oily brown bracelets and anklets which the native craftsmen turned using pole lathes from the Kimmeridge shales of Dorset. But they must surely have considered the Celtic forays into stone carving barbaric – not least the head cult. (Known as enthusiastic head-hunters, the Celts of Britain, both before and during the Roman occupation, carved grotesque stone heads, some of which seem to represent their deities. The cult of the head may predate the Celtic settlement, for chalk blocks carved into drum shapes and decorated with highly stylized faces were made at the start of the Bronze Age.) Troops from the Mediterranean world, with its deep-rooted urban traditions, will also have found the Celtic approaches to town life curiously primitive. There were no glittering colonnades, no imposing public buildings, little, if any, masonry and scant apparent order in the 'town' – the agglomerations of huts and aristocratic estates which often huddled insecurely behind their embracing earthbanks.

Familiar with the monumental building achievements of the older Mediterranean civilizations of Mycenaeans, Egyptians and Greeks, the

Romans introduced to Britain a fully-fledged stone building capability incomparably more sophisticated than anything which had previously existed in the islands. They had also developed the semi-circular arch, and arches of a Romanesque form were to govern the design of the larger stone buildings of Britain for well over a thousand years after the Roman conquest.

Only small and isolated fragments of the imposing masonry of Roman lowland England have survived the traumas of the Dark Ages and the stone-robbers of the centuries which followed. Probably the most evocative glimpses of the lost civilization are found in the shadow of Bath Abbey, where the Roman baths of *Aquae Sulis* were excavated in the 1880s. The Great Bath, with its stone diving board, is flanked by a reconstructed Classical colonnade. An underground network of channels conveyed water from hot springs to various smaller baths grouped around the Great Bath, and nearby, now lost, was the temple of the cult of Sul Minerva, with which the baths were associated. Something of the atmosphere of a refined and selective club has survived the destruction of centuries. The remains of *Aquae Sulis* are still but a morsel of the whole, and it should be remembered that this town and its public buildings belonged only to the second or third division of the colonial league – *Verulamium* (St Albans) covered ten times its area.

From time to time, archaeologists uncover fragmentary relics which remind us of the splendours of Roman England. This was the case in 1974–5, when sections of the riverside wall of Roman London came to light. Though unexceptional as a piece of architecture, the excavated wall demonstrates the practicality and competence of the Roman builders. Carried upon an unstable gravel subsoil, the wall stood upon closely-packed rows of timber piles, while a compacted raft of chalk, rammed down in a layer between and above the piles, formed a firm foundation for the wall itself. The riverside wall seems to date from the second half of the fourth century and may well have been built to protect London against the sea- and river-borne assaults of Saxon raiders. Appearing at quite a late stage in the Roman occupation, not surprisingly the wall was found to contain re-used blocks of masonry from older Roman buildings, and these blocks provided broad hints of the glories of the Roman city. Some fifty-two blocks of Lincolnshire limestone with relief carvings of figures and geometrical motifs were recovered. After long and careful study, experts concluded that the majority of the blocks had formed part of the monumental 'London Arch'. Standing around thirty feet in height, the arch was flanked by larger-than-life figures of Hercules on the front and Minerva on the back, their partners on each side remaining undiscovered. A frieze consisting of the busts of gods, including Venus, Apollo, Mercury and Mars, was carried above the archway.

The London Arch does not appear to have been a triumphal arch, since it lacked reliefs of soldiers, trophies and captives; it may have formed a monumental gateway in a section of the town wall, or have glorified the approaches to an important temple, baths or other majestic public building.

As the efforts to reconstruct the jigsaw of the carved blocks progressed, it became apparent that a group of them must have been pillaged from another monument, which was given the name 'Screen of Gods'. The screen, which would have been at least twenty feet in length, must have formed part of a larger monument. It consisted of the figures of major divinities on the front and representations of mythological creatures and lesser divinities on the back, with the figures standing in niches flanked by pilasters. Also found in the riverside wall were two altars with inscriptions commemorating the restoration of the temples of Isis and Jupiter, and an unusual limestone relief portraying a group of four mother-goddess figures. From such imposing carvings, unceremoniously incorporated into the riverside defences of the Roman capital, we gain the most fleeting glimpse of the achievements of masons and builders in this island outpost of the Empire.

The most complete relics of the Roman occupation are of a military nature, constructed to impress by their robust impregnability rather than ornate sophistication. The most celebrated of the Roman monuments in Britain, of course, is Hadrian's Wall, built at the behest of the Emperor Hadrian who visited the province in AD 122. The seventy-three miles of wall were completed in AD 128, tidying up the intolerable raggedness of the northern frontier which had resulted from the abandonment of Inchtuthil fortress in Perthshire in the closing years of the first century. Although a west-central section of the wall was initially built of turves owing to difficulties in obtaining sufficient supplies of limestone on the west of the North Tyne valley, the wall was able to take advantage of a natural and sometimes cliff-like barrier formed by the Whin Sill.

The blue-black rocks of the sill resemble a volcanic basalt, but their rather larger crystals show that they did not cool as rapidly as in a surface lava flow, but more slowly underground. They are therefore a form of 'dolerite'; at some stage in the Carboniferous period – the era when most of the coal measures and the grits and limestones of the Pennines were laid down – a great sheet, or 'sill', of molten rock around 100 to 200 feet thick was injected into a zone of weakness between the neighbouring beds of shales and sandstones. Ultimately, the sill was exposed through the erosion of the rocks above to form a scarp-like and conveniently north-facing ridge: an irresistible platform for the Roman wall.

The construction of the wall will have been accompanied by a frenzy of quarrying and stone-dressing activity, much of the work being accomplished by the legionaries, who were accustomed to service in road-making and camp-building operations. Many of the quarries used can still be traced, and some near the wall fort of Birdoswald contain crude inscriptions left by the

Roman masonry in an entrance to Housesteads fort on Hadrian's Wall.

stone-dressers. In its western section, the wall incorporates blocks of an unusual rock known as 'brockram', composed of angular fragments of limestone eroded from their parent rocks during flash floods and then embedded in a matrix of red desert sandstone.

In the construction of their southern coastal fortresses such as Pevensey in Sussex, Portchester in Hampshire and Burgh in East Anglia, built to secure the colony against the increasing pressures of continental barbarian raids, the Romans were obliged to make the best use of the available materials. Flint nodules could conveniently be used in the construction of walls, and these have endured since the days of the Saxon threat. They derive their strength from the quality of the mortar in which the flints are set and from the inclusion, every two feet or so, of a horizontal bonding course of narrrow layers of brick. The Roman bricks are much more slender than modern varieties, and normally a rich red in colour. At both Burgh Castle and Portchester, one can see how sections of the walls were built by different gangs, with the 'gang junctions' apparent as irregularities in the horizontal brick bonding courses.

Although the majority of their suburban and villa dwellings will have

Hadrian's Wall runs along the crest of a cliff-faced dolerite sill; this is the scene at Peel Crags.

been constructed, economically, of timber-framing, wattle and daub or brick, and capped with thatch or clay roofing tiles, the Romans were well able to appreciate the qualities of different building stones, and to accomplish the long-distance transport of the more prestigious types – as is the case with the Lincolnshire limestone which was conveyed via the Car Dyke canal for use at *Verulamium.*

A quite amazing capability for the trans-shipment of luxurious stones emerged in the excavation of the palace of a Romano-British noble – probably the client king Cogidubnus – at Fishbourne, near Chichester. The bulk of the walls, which were plastered and exquisitely decorated by imported craftsmen from the Continent, were of local sandstones and limestones, while a variety of hard rocks originating in Cornwall, Jersey and Brittany were imported for use in fittings in the earlier phases of the palace's construction, perhaps arriving as ships' ballast. In the course of the second building phase, white limestones from the Mediterranean area and others of a creamy buff colour from Caen in France were assembled for use in columns; dark red or greenish Purbeck 'marble', true marbles from Turkey, Greece and Spain, and breccias from France were imported for decorative inlays; and sandstones from the Weald were carved into gutter blocks. A few decades

A closer view of Roman flint and brick wall construction from Burgh Castle shore fort.

The ruined defences of the Roman fort of the Saxon shore at Richborough, built of alternating bands of flint and tile-like brick.

before the palace was destroyed by fire (around the 280s), new mosaic floors in multi-coloured imported marbles were laid above the older mosaics.

The collapse of imperial power in Britain early in the fifth century removed the last vestiges of the stability and organization which were essential to large building operations. The old colony was divided and contested by the native, partly-Romanized and partly Christian aristocrats, the British of the provinces, bands of Saxon mercenaries, and Saxon settlers. While the Saxons were vastly outnumbered by the indigenous British, the reins of power were held, in the main, in pagan Saxon hands. Meanwhile, monastic Christianity endured in Ireland, survived in Wales, and filtered into the Pictish territories of northern Scotland.

The urban splendours of Roman England crumbled, although the vestiges of urban life probably persisted precariously in a number of towns. Thus, the bitter contemporary accounts of the Celtic monk Gildas may be somewhat one-sided when he tells of '. . . the foundation-stones of high walls and towers that had been torn from their lofty base . . .' and how 'All the major towns were laid low by the repeated battering of enemy rams; laid low, too, all the inhabitants – church leaders, priests and people alike, as the swords glinted all around and the flames crackled.' Far more than any other inspiration, it was Christianity which stimulated the art of stone carving and building in the western uplands of Britain and Ireland, and which eventually revived it in the English lowlands. The greater ecclesiastical centres, with their ability to organize and control, may have acted as focal points for the distribution of stone for church building in the later Saxon centuries, with York perhaps providing stone for new churches in the surrounding localities. Some Roman stones and bricks were re-used to supplement this supply. The early stone-built minster churches are often associated with royal estate and administrative centres.

While mainland Britain was locked in the turbulent obscurity of its Dark Ages, Ireland, which had escaped both the Roman and Saxon conquests, developed its own distinctive Christian civilization and its own tradition of ecclesiastical architecture. A Christian bishop, Palladius, arrived from Rome in 431, and the British missionary St Patrick appeared about 450. Christianity was soon to proclaim its presence in the landscape through a remarkable series of carved stone crosses, although the roots of this monumental movement lay partly in the enthusiasm for carved symbols that already existed in the pagan land. Perhaps they are even to be found in the peculiar Iron Age carved stones like the Turoe Stone in Galway, the Killycluggin Stone in Cavan, or the Castlestrange Stone in Roscommon: rounded boulders which are confidently and fluently decorated with the incised or relief carvings of curve and spiral motifs so characteristic of the later Celtic settlers in Britain.

An early Christian tomb-slab from the monastic centre of Clonmacnois in County Offaly. Most of the sculpture at monastic sites in Ireland was accomplished on sandstone.

The gap between pagan Irish prehistory and the Christian historic period is straddled by the appearance of stones and pillars carrying inscriptions in the 'Ogam' script, a much adapted form of the Latin alphabet based on groups of strokes which represent letters and lie above, below or through a baseline. Most of the inscribed stones were erected as personal memorials and carry the name of the dead person and brief details of his lineage. The alphabet continued to be used until the seventh century, though in the meantime many of the pagan Ogam stones had been Christianized through the addition of a cross symbol.

While in England the pagan Saxon settlers and war-bands were neglecting or destroying the monuments to Roman civilization and beliefs, in Ireland, Christian monuments in the forms of pillars, slabs and grave slabs bearing the cross began to proliferate. At first, the cross symbols were simple, using the Latin, Greek, 'intersecting arc' or 'wedge-ended' forms of cross. Sometimes, too, a scattering of quartz pebbles around the cross base proclaims the survival of a lost prehistoric symbolism. As the tradition evolved, so the carvings became more elaborate: the intersections of the cross were enclosed within a ring, while figures or elaborate geometrical patterns began to accompany the cross on its upright slab.

Far more puzzling are the *bullauns* which are often associated with early Christian sites. These are rounded pits, sometimes more than a foot in diameter, which are hollowed, singly or in small groups, into the surface of boulders. Many of the chosen boulders are of quartz conglomerate rock. It has been suggested that the *bullauns* may have been used as querns for grinding corn, or as mortars in the preparation of ritual foods for pagan ceremonies. They may then have been adopted by monastic communities and used in the preparation of herbal medicines.

Around the start of the ninth century, the celebrated free-standing carved crosses began to appear, superseding the crosses carved on the faces of slabs and pillars. In some cases these seem to have been manufactured at workshops which had an older tradition of creating carved and inlaid portable wooden crosses: the circle which generally rings the cross is superfluous to a sandstone cross, but would have reinforced the structure of one made of wood. More than a hundred of these stone crosses have survived in Ireland, some only in fragmentary form. A few are undecorated. Most carry a mixture of ornamental and pictorial carvings, scenes from the scriptures, symbolic scenarios which cannot always be interpreted, and the mythical beasts, monsters and ornately interlaced motifs which feature in the Dark Age art of much of western Europe. Originally, the figures and motifs which crowd and tangle on the crosses were probably picked out in brilliantly painted colours. The examples at Monasterboice in Louth and at Clonmacnois in County Offaly, as well as several others, draw scores of admirers.

A detail from the cross at Monasterboice which commemorates Abbot Muiredach; it may date from around 840, or from the early years of the following century.

Church organization on the Roman episcopal model, which the early missionaries had introduced in Ireland, did not take root – largely because of the dispersion of the livestock-rearing communities in small *clachans* or hamlets and the absence of urban foci. Instead, Ireland gave birth to monasticism in the form of a multitude of independent communities scattered about the countryside in sites often chosen for their austerity and solitude. These early Christian communities are not to be compared with the magnificent and opulent foundations which were to be established in Britain during the medieval period. The simplicity and humility which were at first central to the Irish concept of monasticism are expressed in the surviving relics of the age. Where timber was used in the construction of a church and the domed monastic dwellings which clustered loosely within its stockade or compound, all that survives may be the graves and cross slabs of stone.

Some Irish stone churches which seem to be contemporary with the Middle and Late Saxon periods in England have survived in areas where building stone was available as an attractive alternative to timber. Although a knowledge of the use of mortar had been introduced to Ireland by the eighth century, it was only under Norman influence that designs based upon the grander European Romanesque church forms took root there. The dating of early churches is made difficult by the long survival of the indigenous designs, which employed drystone walling in tiny churches lacking in aisles, towers or chancels. In the stone-strewn west, where small religious communities followed hermit-like existences inspired by Middle Eastern examples, a number of tiny churches known as 'oratories' were built. Perhaps evolving from the completely stone-built and beehive-shaped huts, or '*clochans*', which had been built within the embracing ramparts of stone forts and which were still favoured by the monks, the stone oratory was a simple, attractive and ingenious building. The form adopted often resembled that of an inverted boat, with the 'battered' or inward-sloping walls solving the roofing problem and helping to stabilize the windowless drystone walls.

During the eighth and ninth centuries, when Christianity was becoming securely re-established in England, Ireland had become a land of affluent monastic settlements, its calm and confidence being but periodically ruffled by ecclesiastical and dynastic rivalries. Then, both England and Ireland fell prey to the pagan brutality of the Vikings. The remarkable, needle-like stone round towers of Ireland are roughly contemporary with the Norse threat. They are associated with the monastic settlements which will have contained the richest pickings, and they are generally thought to have been built for defence. Even so, these almost uniquely Irish towering cylinders of mortared stonework may have had several functions. Although some have their entrances set defensively between four and fifteen feet above ground level; with around half-a-dozen tiers of wooden floors stacked in the chimney-like towers, their inmates would have had much to fear in the

An early Christian round tower at Turlough in County Mayo.

event of a serious incendiary attack. Moreover, standing like petrified space rockets sometimes over a hundred feet in height, the towers were hardly discreet bolt-holes which invaders might fail to notice! Some of the round towers may have been begun before the appearance of the Norse raiders, perhaps inspired by bell towers seen on pilgrimages to Italy. In addition to their use in broadcasting the ringing of handbells, the towers must have provided the associated monasteries with a certain status; and perhaps their ascent imbued the monks with a sense of a closer communion with God. Whatever their functions, the round towers reveal a very considerable competence in building techniques, which is made more impressive by the apparent lack of antecedents.

In Scotland, meanwhile, little if any masonry was produced to surpass the old Iron Age 'wheelhouse' dwellings, in which the radial partition walls helped to support the corbelled roof, or to exceed the mastery of drystone walling displayed in the northern brochs. The Celtic tribal territories, which ultimately resisted the Roman military advances, then experienced pressures from the Saxons of Northumbria and the Scots from Ireland, who established their footholds in the south-west. Until it was captured and swiftly surrendered its identity to the Scottish dynasty of Kenneth MacAlpin in the mid-ninth century, the Pictish province of the north-east followed an individual, though deeply mysterious existence. Pictish culture and its enigmas are most forcefully evident in the legacy of symbolically-carved stones, which constitute virtually the only obvious survivals of a strange and idiosyncratic nation.

The carvings on the Pictish stones include weird, elaborate symbols of a superficially abstract form, like the 'crescent-and-V-rod', the 'Z-rod' and the 'spectacles'. Gracefully economical portrayals of native animals like the salmon, eagle or deer, human figures, and crosses with ornate interlace decorations in fine relief are also found. It has long since been suggested that the abstract symbols appeared first, date from around the seventh century, and belong to the pagan culture; that the symbol stones which carry both the symbols and the cross represent an intermediate stage in the conversion of Pictland; and that those which carry the cross alone date from the dying decades of Pictish independence. This is quite convincing, but it solves neither the riddle of the meaning of the strange symbols, nor that of their origins. They appear in a fully refined form, and so one can only assume that the symbolism was

One of the Pictish stones displayed at Aberlemno near Brechin, showing the serpent, 'double disc and Z-rod', and mirror case symbols.

developed in perishable media – perhaps metalwork, wood-carving or body tattoos. Whether the symbol stones are territorial markers which sport the emblems of a tribe, or personal memorials with a message about status and lineage, and whether they carry messages of guidance, warning or boastfulness is quite unknown. They present the ultimate challenge to cryptography.

Perhaps because they commonly exist as fragments surrounded by the more elaborate masonry of Gothic styles of building, Saxon churches in England often seem to exude a homely and rustic air. This must be quite unintentional, for the stone-built Saxon church was an exceptional and costly building, sometimes provided by a monastery, but more usually erected at considerable expense by an aristocrat or king. A majority of Saxon churches would have been built of timber, and of these only one example has survived. (The much modified and adapted building at Greensted-juxta-Ongar in Essex still retains its walls of close-set vertical oak logs and, with a measure of imagination, enables us to visualize a type of church which was probably common in the Saxon landscape.) The other surviving Saxon churches are stone-built, and in this respect far more durable and distinguished than the lost timber halls or thatched huts of their creators and congregations. In many places, the funds needed to provide even a timber church were at first beyond the means of the villages, hamlets and their overlords, and the church was preceded by a preaching cross erected at a convenient meeting place or a spot held sacred to an older pagan tradition.

In some places, the broken remains of a Saxon cross can be seen incorporated into the masonry of a later church. Occasionally, its fragments come to light in the course of a church site excavation, and a number of tumbled crosses and cross shafts have been re-erected. A fine pair of shafts are displayed in the market square at Sandbach in Cheshire, a trio grace a churchyard at Ilkley in Yorkshire, and the churchyard at Gosforth in Cumbria displays a lofty Late Saxon example, some fifteen feet in height and complete with its ringed cross head. Although experts can differentiate between the products of the different national and regional schools, Dark Age art and stone carving, with its intricately interlaced decoration and lively, energetic beasts contrasting with the stiffness of the human figures, spans the cultural divides.

Sometimes, the remains of Roman buildings will have suggested ideas to the Saxon masons as well as providing convenient quarries, while other, more complete Roman monuments will have been seen by churchmen in the course of continental training and pilgrimage. The manner of building which developed belonged to the Romanesque tradition: the round arch was a key component of the Saxon church. But the churches of this period are quite distinctive, embodying some features which derive from distant

The Saxon church at Breamore in Hampshire is built of flint; stone has been imported to form the quoins.

Classical inspirations and others which must have been adopted as the masons gained experience. The Saxon churches lack both the decorative and constructional sophistications of their post-Romanesque successors and the symmetry and finish of the earlier Classical buildings, and their very considerable attractions often derive from the rough-and-ready nature of the stonework. It would be quite wrong to judge the Christian civilization

The Saxon church at Escomb employs much re-used Roman material, possibly including the chancel arch.

which developed in England in the centuries following the Saxon conversion from the fragmentary evidence of the surviving churches. Though class-ridden, riddled by aristocratic intrigues and severely afflicted by the Danish invasions, Saxon society was in many respects more cultivated and civilized than that of the Norman conquerors. In the fields of literature, manuscripts, needlework, and probably the construction of timber buildings, the nation was unsurpassed.

The craft of masonry in Saxon England was almost exclusively confined to the building of churches. By virtue of their great antiquity, and of the fact that the Norman monarchs seem to have gone far beyond the need to renovate or extend in their destruction and replacement of Saxon churches, only fragments of the Saxon legacy of churches survive. We do not know how representative these fragments are. It is not accidental that many of the most celebrated survivors – like Bradford on Avon, Brixworth, Earls Barton and Brigstock – are associated with the excellent and easily-worked outcrops of the Jurassic limestone belt. In a number of cases, Roman bricks were incorporated into the church masonry. At Escomb near Bishop Auckland, perhaps the best preserved of the northern Saxon churches, many of the stones display Roman tooling and were probably robbed from the ruins of the fort at Binchester – the chancel arch itself seems to have been re-cut from a Roman arch. We even know that some of the re-used stones were first quarried and dressed by the troopers of the Sixth Legion, for one of the stones in the nave is inscribed 'VI LEG'. In East Anglia, where only flints and the occasional bed of 'conglomerate' or puddingstone were to hand, the circular church tower was developed to obviate the impracticalities of forming a corner from flint nodules. The abundance of these round towers suggests that they may have been built partly for defence.

In order to understand the aspects of masonry which are distinctively Saxon, we can do no better than to examine the beautiful tower of Earls Barton church in Northamptonshire. The battlements and the main body of the church belong to various medieval periods, but the Saxon church was very much shorter, and was completely dominated by its tenth-century tower. The village name probably means 'The grange of the earl', and it may be that the noble who caused the church to be built was not oblivious to the defensive potentialities of its majestic tower, built upon a natural spur which is enhanced by the earthworks of what may be an Iron Age promontory fort. From the photograph, it can be seen that while the bulk of the tower masonry is rendered, narrow strips of stone project to form vertical and X or lozenge-shaped patterns. This 'raised flat band' form of decoration was favoured in the later Saxon period, though whether it attempts to mimic the framing patterns upon contemporary timber buildings, we do not know.

In any rectangular stone building, the most severe stresses are likely to be concentrated in the corner angles, and the Saxons were almost invariably distrustful of their ability to form corners without resort to massive reinforcements. Quoins of 'long and short work', in which upright pillar stones alternate with other large blocks of masonry, set horizontally and biting deeply into the surrounding stonework, are normal in Saxon churches – and are clearly displayed in the quoins of Earls Barton tower.

Saxon window openings sometimes had round and sometimes triangular heads: both types may be seen at Earls Barton. Frequently, pairs of long, narrow, round-headed windows were joined by a small stone column or 'baluster', rounded and sometimes shaped in the manner of an old lathe-turned chair leg. At Earls Barton, round-headed windows separated by baluster shafts occur in groups of five on the different faces of the upper section of the Saxon tower. Like their arches and most of their window openings, Saxon doorways were normally round-headed. A fine example with a moulded arch forms the main entrance on the west face of the tower, and a second doorway, with its sill resting on the projecting 'string course' which defines the base of the second stage of the tower, is on its south face, some twenty-two feet above the ground.

Most of the characteristically Saxon features which are evident at Earls Barton are repeated in other Saxon churches, so that these buildings are distinctive and readily recognized even when they remain only as fragments within a later building. The exception is Brixworth in Northamptonshire. To visitors accustomed to the delightful simplicity of the typical Saxon church, Brixworth will come as quite a shock. Described by Sir Alfred Clapham as 'Perhaps the most imposing architectural memorial of the seventh century surviving north of the Alps', the church at Brixworth is remarkable not only for its size and near completeness, but also for its very early date. It belongs to an era of missionary activity, and may have been built by monks who were seeking to convert the pagans of Mercia. The amber-tinted stone of the church is echoed in the stone cottages of the village below, and close inspection shows that this is a tufa limestone, which is formed by the precipitation of calcareous material from running water (as can be seen on the beds of petrifying streams). Much more obvious is the use that has been made of Roman clay tiles, which must have been pillaged from the ruins of neighbouring villas. Laid with no great regard for precision, the rust-red bricks have been used for functional and decorative effects to form round arches above the doors and windows. Another not uncommon Saxon building device, which is displayed in the projecting stair tower at Brixworth and in the walls of the churches at Deerhurst near Gloucester and Diddlebury in Shropshire, is the laying of narrow slabs of stone rubble in herringbone courses. In addition to its decorative effects, the herringbone method allowed courses of an even height to be laid, irrespective of the thicknesses (but not the lengths) of the slabs used.

In the course of the Bronze Age, interest in the construction of imposing Megalithic monuments waned. The more prosaic craft of drystone walling continued to be practised, but it is unlikely that there was any great progress or interest in the construction of stone buildings in the millenium which preceded the Roman occupation. Although the splendid Roman architectural creations were robbed of stones, neglected or decayed during the centuries following the collapse of the Empire, a residue of knowledge and skill survived in Britain and western Europe. The revival of Romanesque building traditions in the construction of stone churches in England, France and Normandy provided an essential stage in the progress towards the splendours of medieval Gothic architecture.

Chapter 11

Springtime and High Summer

*A*s Alec Clifton-Taylor has pointed out, '. . . in every branch of art except architecture, the Norman Conquest actually resulted, for a while at least, in a decline from the achievements of the Anglo-Saxon period.' Even in the field of architecture, there may be question marks, for much of the earlier Norman work in England was poorly built, uncertainly surveyed, prone to fall down and crudely decorated. The Norman architectural style was another offshoot of the Romanesque tradition and, if the evidence of the Abbey of Edward the Confessor at Westminster as depicted in the Bayeux Tapestry is to be believed, the Norman style may have gained a small foothold in England before the Conquest.

The English nation, which was beginning to form from the integration of its British, Saxon and Scandinavian parts, fell into the grasp of the Bastard of Normandy and his army of part-civilized Frenchified Vikings and opportunist fellow travellers. What the invaders lacked in civic visions or interest in the common weal, they made up for by their amazing energy. Following a brief interlude, in which their political triumph was consolidated by a comprehensive programme of earthen castle building and during which continental monastic orders were attracted here by the complexion of the authoritarian but aggressively 'Christian' realm, an unprecedented campaign of church building in stone was launched. Despite his boorishness – his passion for bloodsports, his undisguised contempt for inferiors and 'inferior' cultures, his enthusiasm for defence, blindness to issues of social justice, his greed, and his deep conservatism – William and his equally unattractive offspring were to transform the face of England with their hundreds of churches and castles and, through conquest or influence, also affect the landscapes of Wales, Ireland and Scotland. But the gross, monolithic qualities of some Norman architecture remind one of the Nazi and Stalinist extravaganzas of Central and Eastern Europe. It announces, with a brutal stridency, that the old order has changed and tough new hands are grasping the tillers of society.

At the start of the twelfth century, a staggering programme of works was already in hand, with massive new cathedrals rising like noble galleons above seas of stacked timbers, stone dumps and foundation trenches, while small armies of masons, carpenters, smiths and labourers scurried all around. At St Albans and Winchester, Durham and York, Peterborough, Chichester and Ely, the scenes were repeated. Both the organization and the competence of the Norman administration have been exaggerated – it is the Normans' vigour which enthrals. While the cathedrals which erupted from the many English launching pads dwarfed their older Saxon counterparts – like the rough-walled conglomerate (presumed) cathedral at North Elham in Norfolk, which was hardly more than 130 feet in length – a thorough-going parish church-building venture was also in hand.

We do not know how many English stone churches existed at the time of the Norman Conquest, but the evidence of fragments of crosses and grave slabs, broken arches and church fittings, which can be found built into the walls or buried in the vicinity of medieval churches implies that there were far more of them than was once thought. By the time that Domesday Book was compiled in 1086, there may have been as many parish churches as could be found in the much more populous nation in the middle of the eighteenth century, on the eve of the Industrial Revolution. Many of these churches disappeared completely in the course of successive reconstructions, while others served villages destined to be 'lost' by the time the medieval period had run its course. By the end of the eleventh century, Lincoln, an important provincial centre but only a fraction of the size of the modern town, supported some thirty-nine churches. The Norman ecclesiastical building boom at first had its energies chanelled into the building of cathedrals. Some decades passed before there was a wholesale building and rebuilding of parish churches, so that the churches built in the latter part of the eleventh century, seemed as much, or more, Saxon as Norman.

With possible minor interruptions, as in the reign of King John (1199–1216) when the realm lay under a papal interdict, the pace which was set in the era of the Norman kings was maintained. As well as founding new churches, the Normans were apparently zealous in substituting their own distinctive buildings for those of the Saxons. Many of the new Norman buildings, however, were too poorly constructed to survive the passage of time. The Norman towers of both Hereford and Ely cathedrals, for example, collapsed. The most dynamic period of parish church building and reconstruction probably occurred between 1150 and 1250. By the middle of the thirteenth century, there seem to have been ample churches to go round, and so most of the subsequent work involved repairing the existing churches

Norman architecture at its most imposing can be explored at Durham Cathedral: the nave (above); the entrance to the cathedral (right).

or adapting and extending them to follow new fashions and meet new demands. On the eve of the Great Pestilence, which arrived in 1348, London (which might then have compared in size to a fairly compact modern regional centre like Harrogate in Yorkshire) boasted more than a hundred churches. As W. G. Hoskins has pointed out, even Rutland, a lost county which is studded with the overgrown remains of lost villages, had more than fifty medieval churches contained within its modest bounds.

The ecclesiastical building outburst was by no means confined to the construction of parish churches. D. Knowles and R. N. Hadcock have calculated that in 1066, the monastic population of England may have numbered little more than 1,000; by 1217 this population had multiplied twelvefold. Continental monastic orders were attracted and courted by the Norman overlords, though the Benedictines were well established before the Conquest. Whether they were baited with generous grants of lands or began their occupation of the countryside in conditions of humble subsistence, endowments soon flowed in, and the estates were harnessed into rich production. Early in the thirteenth century, the monastic communities, which were divided between some 180 large foundations and about five hundred lesser houses, controlled perhaps one-fifth of the national wealth. Just before the outbreaks of the Pestilence, which were to cut the population to around two-thirds of its former size, the kingdom contained around 17,500 monks, along with hosts of lay brethren and monastic tenants and servants. A number of the early cathedrals, such as Ely, Durham, Canterbury and Norwich, were also monasteries of the Benedictine order (whose rules had been adopted in English foundations in the tenth century), while the great northern monasteries, like Fountains, Rievaulx and Meaux, were established in the mid-twelfth century by Cistercians, in lands still sterilized by the Conqueror's Harrying of the North.

Richard Morris has studied the rates of medieval building, and describes how around thirty cathedrals and important monastic churches were begun in England between 1070 and 1100, over half of them being started in the crucial decade 1090–1100. At the same time, the kingdom was sustaining a great programme of castle building, so that during this decade work was probably begun or in progress on around fifty major cathedrals, abbeys and fortifications. Professor R. Allen Brown suggests that at least fifty castles had been built by the end of the Conqueror's reign in 1087, and by the end of the eleventh century, the figure was at least eighty-four. He points out, however, that these are examples specifically mentioned in the surviving records, and probably just a fraction of the actual total.

We know next to nothing about Saxon stone domestic architecture, and perhaps next to nothing existed. To a considerable extent, the monopoly

One of Lincoln's Jews' Houses, a much-modified Norman building. In medieval England, Jews served as bankers and money-lenders and were periodically the victims of violence and extortion. They could both afford and needed the advantages of life in a stone-walled dwelling.

of ecclesiastical architecture on the use of stone continued in the Norman period, though castle building increasingly made its demands upon the quarries. A number of Norman dwellings were built of stone; few have survived, and it is not easy to judge their original frequency. The famous Jews' Houses in Lincoln are much altered, but preserve some features of the Norman period; the School of Pythagoras in Cambridge is another invaluable example, but the manor house of Boothby Pagnall in Lincolnshire, which dates from about 1200, is the best of the survivors. It is a type of building known as a 'first-floor hall', with the domestic accommodation, a hall and solar, sited on the first floor, above storage rooms, and served by an external staircase. Stone was more suited for such manor houses, as being more prestigious and more easily defended. In the event, these Norman buildings seem to have underlain the development of the later medieval defensive peel towers and castles of the troubled Scottish border zone, but the timber-framed Saxon hall played a much more central role in the evolution of the house. About forty examples of Norman domestic architecture are known – mainly small fragments of the original buildings. As Dr

Margaret Wood has pointed out, most of them owe their survival to the fact that they were in stone. They are closely linked to the leading quarrying districts – notably the Jurassic limestone belt running through Lincolnshire and Northamptonshire to the Cotswolds and Dorset, with outliers in the Millstone Grit country of Yorkshire and flint-strewn chalklands of the southeast. More stone-built halls of the later medieval centuries are known, such as the first-floor halls at Meare in Somerset, Brinsop Hall in Herefordshire, Fyfield Manor in Berkshire, and Yardley Hastings Manor in Northamptonshire, all of the fourteenth century. On the whole, though, while stone supplanted timber in the construction of castles, most manor houses and virtually all inferior dwellings employed one or other of the many variants of the timber-framing tradition.

In the immediate aftermath of the Conquest, the victorious fortune hunters beavered away to secure their new estates, casting up moated and palisaded earth mounds which were sufficient to intimidate the peasantry, but inadequate for sterner defensive duties. The death of the Conqueror, the rise of provincial aristocrats no less rapacious than he, the dynastic intrigues, the Civil Wars of King Stephen, and the refinement of siege warfare all made increased demands on the royal or baronial stronghold. The story of the evolution of the medieval castle is a complicated one, with each successive chapter (as from square to round to D-shaped towers) being necessary to ensure survival in a violent and unstable age. The first essential change involved the substitution of the wooden motte-crowning palisade by a tower keep or curtain walls of stone. Subsequent stages involved the uses of longer, newer and better walls in ever greater thicknesses of masonry – until eventually most, except the royal coastal fortresses, toppled over the cliff of redundancy. As the motte and bailey castle was superseded by the keep and curtain wall designs, the cost of fortification became a heavy burden on those kings and nobles who sought the security of stone walls. It has been estimated that, in the reign of Henry II, the income of the average knight was around £10 to £20 a year, while the average annual revenue of the government of the kingdom was about £10,000. Of this state revenue, at least a tenth was swallowed up by castle works and maintenance (anticipating the modern tendency in defence spending). In some years, almost half of the revenue was consumed in this way. The records are rather sketchy, but particular details have survived concerning the new castle which Henry built at Orford in Suffolk – and which he greatly admired. Professor Allen Brown describes how, between 1165 and 1173 a sum of £1,400 was spent on the castle – and he also notes the contrast between this stone castle, which took eight years to build, and the earlier earthen mottes, which were said to be erected in eight days! At Orford, the builders used the poor local 'septaria', nodules of limey clay and marl which were the best that this stone-poor region could offer. Had a good stone been imported from a source like Caen or Barnack, then Henry's castle would have been more costly still.

In the course of the centuries leading up to 1348, church and cathedral architecture became ever more ornately sophisticated; castles were intended to make a more basic impact. So, while the flourishes and decorative mouldings of the rapidly-evolving craft of stone carving were largely inappropriate to the castle, each new device to thwart the enemy without had to be evaluated and incorporated, while 'gigantomania' was a desirable if costly pursuit. The large castle was no less demanding of rough labour than the cathedral; the rival undertakings would compete for their shares of the labour market, and the royal castle had a distinct advantage over other projects: the king could conscript the necessary artisans from other building sites. As an example of the demands created by the early fortress, D. Knoop and G. P. Jones have quoted the case of Beaumaris Castle, which at one stage in its construction provided work for 400 masons, thirty smiths and carpenters, 1,000 labourers, and a wheeled armada of 200 carters.

Let us take a detailed look at the case of just one great castle – Caernarfon – where the documentary record is more complete than usual. The scale of the undertaking was immense, and might be compared today to the 'Space Race' in the way that the resources of the state were organized to achieve a particular goal. The pacification and colonization of Wales during the reign of Edward I transformed the principality – and of course the work on the castle at Caernarfon was paralleled by other fortress-town constructions taking place simultaneously at Conway, Criccieth, Bere, and Harlech.

The building of a castellated borough at Caernarfon was in part a reaction to the Welsh revolt and Edward's subsequent victory over Llewelyn in 1282. Caernarfon was conceived as a link in the chain of English castles which bridged the gap between Conway and Harlech, and as a secure base from which to control the Menai Straits and the crossing to Anglesey. Work began with the construction of the ditch and town wall. Then work on the castle itself commenced, but the fortress was not completed until 1322. The designer of the castle was James of St George, who, despite his name, was an Italian of considerable genius and 'Master of the King's Works in Wales'. Between 1277 and 1301, work was begun, if not completed, upon eight massive Welsh projects – Flint, Rhuddlan, Builth and Aberystwyth in the first sequence, Conway, Harlech, Caernarfon and Beaumaris in the second. We know the names of some of the other shadowy figures involved in the enterprise. Thomas de Esthall, the Chamberlain of North Wales, was responsible for ensuring the collection of the necessary revenues; and the master mason for much of the time was Walter of Hereford. The costs were enormous. By 1301, with two decades of work still to come, Caernarfon is esti-

The incredibly costly Edwardian castle at Caernarfon.

mated by Professor J. G. Edwards to have cost around £16,000; while in the single year of 1291, the projects at Caernarfon, Conway and Harlech cost £14,000. What these figures represent in modern values is really impossible to calculate, but we would be dealing in many millions of pounds, and it must also be remembered that the raising of state revenue then posed far greater problems than it does today. To put the figures into some sort of perspective, Henry de Elerton, who succeeded to the key post of master mason in 1315, was receiving a wage of four shillings a week during his employment as under-master mason. One of the skilled masons under his control might be earning in the region of 13p.

Whether or not the Welsh workers possessed sufficient skills, they were not trusted to build the Edwardian castles, and the work was largely accomplished by scores of English conscripts. Many, one suspects, were recruited against their wishes, and a good proportion of them will have been press-ganged from other, less strategically important undertakings. Thus, at the various Welsh fortress sites, masons from Yorkshire, Northamptonshire and Dorset will have mixed with carpenters impressed from Derbyshire and ditch-diggers from Lincolnshire: the building sites must have served as melting pots in the development of the national culture. In the summer of 1295, the Justice of Chester was commanded to select 100 masons and to despatch them to Caernarfon carrying their tools. (With the emphasis upon quotas, one wonders whether the Justices and Sheriffs responsible for fulfilling such commands were conscientious in considering the skills of their conscripts.) Once impressed, the masons will have had difficulty in evading their obligations. They will have found themselves allocated to different tasks according to their abilities and had many opportunities to learn from their peers. That great variations in their craftsmanship existed is shown by Knoop and Jones, who point to the seventeen different rates of pay applicable to masons at Caernarfon in 1304.

Whether the end product was a cathedral, large monastery or fortress, the medieval building site was a scene of intense activity. We must imagine not only the permanent building rising slowly from its footings, but also the workshops or lodges erected to serve the separate specialists, the shanty town of temporary accommodation, the dumps of earth, stone and timber, and the swarms of labourers, masons, carpenters, and smiths – all bustling and jostling under the gazes of their respective masters, bureaucrats and paymasters. Such projects will have greatly boosted the local economy and stimulated the local service industries. Anyone possessed of a humble cart or a barge could expect regular contracts, plying back and forth between the building site and the quarries. All other things apart, these gigantic undertakings foreshadowed Keynesian policies, abolishing all unemployment for miles around. There must be lessons here for our present masters.

The number of masons employed at Caernarfon varied according to the stage of the work and the season. It has already been mentioned that orders were given for 100 to be despatched in 1295, during the early stages of the castle building, and perhaps some of these men were employed in quarrying. In 1304, fifty-three masons were at work on the site, and thirty-five quarrymen were employed in the four quarries which furnished the stone for the castle; a dozen years later, the demand for stone was still sufficient to provide work for thirty-three quarrymen. As we shall see, the demarcation between the grades of masons and the quarry workers was not so exact as were the divisions in those trades where the guild system had a greater impact: in the winter of 1316–17, a number of the men who were scappling or rough-shaping stones in the quarries were skilled stone layers despatched from the castle site.

The castle will have become the focus of a network of routeways, churned and rutted by the regular passage of carts carrying sand and lime for use in mortar, timber – unsawn or prepared into boards, beams or studs – and stone. Most of the materials, however, came to the coastal site by ship and barge. Four royal vessels were provided, the crew being paid from the building budget, and other vessels of twelve to twenty tons were chartered. The accounts, for example, note the use of ships belonging to David Da and Simon of Cardigan for the carriage of timbers from Trefiw to Caernarfon. The small army of medieval masons were in the vanguard of the remarkable project, and we know little of the individuals who bustled in the many branches of the back-up services which were essential to support the masons' work.

Edward died in 1307, when the castle was only half finished. We are more fortunate, for we can admire the finished, though never quite completed, product. Thousands do this on every summer's day, but the castle becomes that bit more impressive if the visitor pauses to consider the scale, the scope and the logistical triumph of the task – even though it symbolizes the demise of a nation's independence.

It was in the building of the church, the cathedral and the new monasteries that the techniques of masonry were developed and refined. Just as the Saxon church has its distinctive features, so too it is hard to mistake a Norman building. Much Norman work has a heavy, rather oppressive air. The church walls are generally massive, up to four feet in thickness, with the different bays often defined by flat, shallow buttresses which seem to be more embellishments than effective supports. The apparent sturdiness of these walls may mask a lack of confidence, and also of competence, on the part of their builders: many of the earlier churches will have been built by untrained peasant conscripts labouring under threats of violence. Difficulties associated with implementing even the simplest of church plans may be discerned; the mortar used is often of a poor quality, and the failure to effect a strong bond between the outer masonry of the wall and the inner rubble core was a frequent cause of collapse.

In lesser churches, the masonry was usually composed of a 'coursed rubble' of small stones, and gaping chinks between these stones are often seen. Where appearances were held more important and a 'freestone' which could be sawn to shape was available, then an ashlar facing of smoothed and squared stones might be provided, to mask the sometimes unstable rubble core. The same technique was commonly used in the building of columns, which were smoothed and rounded without, but filled with a rubble core. When the walls were built of flint rubble, squared stones of a better quality were needed in the construction of the quoins. At first, the Normans were prepared to import the attractive oolitic limestone of Caen from their homeland. In due course, English freestone quarries in the Jurassic limestone belt were more fully developed at places such as Barnack, now in Cambridgeshire, where the workings were managed by Peterborough Abbey, as has been described.

The older Norman churches tend to be plain, claustrophobic and austere. The mouldings surrounding the arches are simple, the windows often mere slits, and the 'cushion' capitals which surmount the columns are cut from a single cube of stone, their lower corners rounded off to meet the junction with the cylindrical column. After the close of the eleventh century, the decoration displayed on the masonry became more evident and more flamboyant, and at the end of the Norman building phase it was sometimes even florid. It is generally most obvious in the mouldings surrounding the arches of the chancel, nave arcade and doorways. The unsophisticated geometrical designs were at first based upon the zigzag or chevron motif, unceremoniously hacked into the stonework with the axe rather than the chisel. In due course, a richer decorative repertoire evolved: the beak-head, pellet and lozenge forms were among the many motifs used. Gradually, the axe yielded to the chisel, which, in the right hands, could be used to carve swirling curves, sharply and deeply undercut. Thus was the way prepared for the development of a European Gothic tradition which was neither based upon the recreation of Roman forms, nor so reminiscent of the rich decorations of the eastern world. Nobody who has visited the unusually ornate little Norman churches at Barfreston in Kent or Kilpeck in Herefordshire can return entirely ill-disposed towards Norman work.

The larger stone buildings of the early medieval period were constructed as shells of shaped freestone, which were filled with a packing of rubble. At Bury St Edmunds Abbey, the shell has decayed, revealing the rubble cores.

In the course of the Norman ascendancy, new quarries were opened, and others which had been worked by Romans or Saxons were revived or expanded. Even so, with only carts, rather primitive wagons and narrow river barges to move cargoes of stone across a kingdom whose road network was still dominated by the poorly-maintained legacies of the Roman Empire, transport was a major problem. The costs involved formed a substantial component in the building accounts of all important medieval constructions, and the restrictions endured until the days of the canal and the railway – by which time the greatest achievements of masonry had already been accomplished. The expenses incurred in the movement of stone were often considerably higher than those of buying or quarrying stone at its source. The accounts relating to the construction of York Minster have been explored by Knoop and Jones, and the transport costs in this case seem to

Norman carving at its most vigorous and fluent, incorporating a range of imported motifs in the doorway of the parish church of Kilpeck near Hereford.

be fifty per cent greater than those of quarrying. From the quarries at Thevesdale and Bramham, the stone was carted to the river port of Tadcaster; from Stapleton quarry it was taken to Wheldrake port; and from the famous Magnesian Limestone quarry at Huddleston it was transported to Cawood port. From these river ports, the stones travelled by barge to York and then from the wharves by sled to the Minster site. In areas where river barges could not operate, the costs of transporting stone over even moderate distances were likely to be considerably higher.

Since at least Roman times there had been a general awareness that certain building stones were far superior to others. Near perfection was offered by the quarries of the Jurassic limestone belt (see pages 171–3). Amongst the medieval cathedrals and abbeys which exploited these limestones were Bristol, Gloucester, Lincoln, Oxford, Peterborough, Salisbury and Wells. It is notable that all these churches had good access to the limestone belt, while York Minster, as we have seen, was served by nearby quarries with the near-comparable Magnesian Limestone, and Winchester imported limestone from the Isle of Wight. In several other cases, the barriers imposed by distance were too great to be overcome, and other sources of stone were exploited. Durham Cathedral and Kirkstall Abbey were built of a quite acceptable though less easily worked local gritstone, though at Chester and Carlisle in the north-west and Lichfield and Worcester in the West Midlands, builders resorted to the more easily eroded wares of New Red Sandstone beds.

However, some of the later movements of stone were staggering, and outstripped even the operations that shifted Caen stone to southern England. In the latter part of the Middle Ages, King's College, Cambridge, incorporated stone which was quarried near York and moved by barges, North Sea shipping and Fenland river boats to its destination beside the River Cam. Much later, in the 1720s, the renowned sandstones of the Aislaby quarries near Whitby were shipped to King's Lynn and then hauled to the building works at Houghton Hall in Norfolk. The Barnack quarries, first exploited by the Romans, dispersed their wares widely throughout East Anglia during the medieval period. Barnack was well served by nearby rivers – the Welland and the Nene – and by Roman canals like Car Dyke and Foss Dyke, while the Clipsham quarries had access to a tributary of the Welland, the Glen. Some waterways retained their importance, and Alec Clifton Taylor and A. S. Ireson record that, in the middle years of the nineteenth century, the annual value of stone carried on Welland barges averaged around £10,000. For short hauls in the Middle Ages, stone could be carried on low-slung waggons or sledges, though the nature of the local terrain exerted strong controls over land transport.

The first of the new Norman cathedrals was built at St Albans, where the shadowy figure of the martyred legionary St Alban was the focus of a considerable cult. Had he had the foresight to be martyred at Northampton, Bath or Oxford, then the building problems would have been much less, but Hertfordshire lacks building stones. The builders were, therefore, obliged to work with the nodules of flint which were liberally scattered in the chalky fields, while the tumbled debris of Roman *Verulamium* surrendered cartloads of Roman bricks. Despite the vulgarity of the Victorian alterations, the cathedral, which became the premier abbey of England and stayed so until 1396, is a striking and colourful building, with rust-red bricks predominating over the white and grey stipple and banding of flint.

There is very little detailed documentation to describe the building works in Norman England. One exception, however, concerns the account by the monk Gervase which relates to the rebuilding necessitated by the conflagration which consumed the choir of Canterbury Cathedral in 1174. Following a competition, the work was entrusted to a Frenchman, William of Sens. Stone was shipped in from Caen and machines were devised for the unloading of stone at the docks and its haulage to the building site; meanwhile,

The Norman cathedral at St Albans incorporates masses of bricks which were pillaged from the ruins of nearby Roman *Verulamium*.

templates showing the profiles of the chosen designs and mouldings were delivered to the assembled masons. After the work had progressed for four years, William was working on the great vault when scaffolding or machinery snapped beneath his weight and he fell some fifty feet amongst an avalance of beams and masonry. For a while, he continued to direct operations from his sickbed, until worsening health caused his departure to France; the task was then entrusted to an English mason.

In the course of the twelfth century, the church and cathedral builders pursued their tasks with vigour, undaunted by failures and disasters, accumulating new insights as they progressed. Their visions, however, were held in check by the limitations of the Romanesque form. The semi-circular arch lay at the core of the problem. It could form the basis of a half-cylindrical, tunnel-like 'barrel vault', which could be of indeterminate length and of a width and height governed by the capabilities of the makers of the arch. However, such barrel vaults were enormously heavy and placed mighty stresses upon the supporting walls and columns, which in turn had to be massive and costly. A square space could be vaulted by forming a pair of barrel vaults which intersected each other at right angles, but the area which could be roofed was still restricted by the weights and stresses which must be borne by the walls; and the essence of the church plan was the long nave – in the smaller churches, commonly twice the length of the chancel.

In Durham Cathedral, the eventual solution to the problem was anticipated in the construction of Europe's first high-rib vault. Begun in about 1110, it had pointed transverse arches above the transepts, and these were soon followed by the revolutionary pointed transverse arches above the nave. However, in other Norman cathedrals, such as Ely and Peterborough, the problem was avoided and the original naves were simply roofed in timber. It was only in the final quarter of the twelfth century that the possibilities foreshadowed at Durham were fully explored with the introduction of the pointed arch. Since the height of an arch could now be governed by the sharpness or shallowness of the pointing, the old restrictions on width and form imposed by the Romanesque arch were removed. The experiments with arch forms also stimulated an awareness that walls and columns need not be uniformly massive.

If the thrusts and stresses could be focused on particular sturdy columns, or upon enlarged and improved buttresses, then the remaining walls and supports could be lighter and more slender. And if a wall, thereby, was not supporting an enormous weight of beams or vaulting, then it could be pierced by larger window spaces. As the interior of the church was, in consequence, more brightly illuminated, more attention could be paid to improving the finish and quality of the internal masonry and painted decoration. The possibilities were profoundly exciting.

Whether the new style found its first English expression in cathedral works at Worcester, Canterbury or York Minster, or whether it was pioneered in humbler parish churches, is a matter of much debate, but we know that pointed arches were a feature of the rebuilding of Canterbury Cathedral in 1175–85. At some other places, more conservative exponents persevered with rounded arches until the close of the century; so the change from the Norman to the first of the Gothic styles was transitional rather than abrupt, and in Ireland it was even more delayed. In a number of cathedrals, such as Ely (in the façade) and lesser churches, such as Castle Hedingham in Essex, one can see pointed arches embraced by mouldings bearing geometrical motifs of Norman character. By 1150, the new 'Early English' style had emerged as a coherent and individual form which contrasted quite markedly

Transitional architectural styles appeared as the Norman architecture began to yield to the Gothic, and is excellently displayed at Ely Cathedral.

with the Norman Romanesque. It developed during the frenzied era of cathedral building and at a time when the still youthful monasteries were beginning to harvest the riches of their vast estates. Not surprisingly, the new style was embodied in the new and the still developing cathedrals and adopted in many of the important monastic buildings – such as Rievaulx Abbey and Fountains Abbey. It provided the inspiration for some of the most breathtaking of frontages, as at Wells, Peterborough, Lincoln (where the

York Minster, a Gothic masterpiece built from the products of several Magnesian Limestone quarries.

Norman core was retained after the cathedral had been demolished by an earth tremor in 1185) and Ripon.

Richard Morris suggests that a relative lull in great building operations in the 1130s and 40s could partly reflect the disruptions caused by the anarchy and disorders of King Stephen's era, but that more probably it relates to the fact that many of the schemes launched in the boom years (1080–1120) were still in progress and not yet ready for large-scale renewal.

In the second half of the twelfth century, the pace accelerated: the Gothic style encouraged new building works in the revolutionary mode, and many of the Cistercian and Augustinian communities which had been established during the second quarter of the century were now aspiring to create impressive monastic churches. In 1208, however, John's realm was placed under a Papal interdict, and ecclesiastical building programmes were often suspended during the troubled decade that followed. Works were then revived and progressed at a rapid rate, and it can be argued that standards of craftsmanship and the quality of stone building came closest to perfection during these last phases of the Early English style. Gothic architecture had captured the imaginations of scores of wealthy patrons. Meanwhile, the growth of the English wool industry nurtured a prosperous church and nobility, and provided the wherewithal to sustain ambitious schemes.

Like all other forms of architecture, the new Gothic style reflected the spirit, conditions and capabilities of the age. In modest churches more than in opulent cathedrals, it expressed the serious, almost severe intensity of a monastic movement which was still not wholly corrupted by success. The ornate cathedrals, on the other hand, reflect the concentrated wealth and power of the church in a realm where the secular powers, the kings, claimants and barons were often in competitive disarray. The oppressive character of much Norman architecture, which often seems to bear down upon the beholder, was replaced by the liberating emphasis of the up-sweeping line. This ethos was conveyed by the slender vertical shafts grouped around the graceful columns, the heavenward pointing arches, and the deeply-fluted mouldings which defined them.

The new buildings also embodied and encouraged the rapidly developing skills of the masons who made them. The ornate carvings of fabulous beasts made on the capitals in the crypt of Canterbury Cathedral around 1120 are a remarkable advance upon the geometrical hatchet work of the mouldings of older Norman churches, and anticipated the splendours which were to follow.

Windows began to present opportunities for elaborate embellishments. The essence of the Early English style was the tall and narrow pointed lancet window. Masons must have experimented with the possibilities of grouping lancets together in pairs or trios, having realized that a single, protective, pointed hood mould could be used to protect the group from dripping rain. They would then have seen that most of the masonry between the lancet peaks and the hood mould was superfluous. By piercing the space with simple openings, they established the basis for window tracery.

The tower, too, began to suggest decorative possibilities. Norman towers had tended to be plain and squat, though sometimes quite richly embellished with blind arcading. They also had a marked tendency to fall down. Ely has what may be the best of Norman façades, but the collapse of the Norman tower provided the opportunity for the construction of the remarkable fourteenth-century octagonal tower, with its cluster of four flanking octagonal turrets. The spire, rather than the tower, became the crowning glory of the Early English church. Obviously, masons whose grandfathers had groped with the problems of tower construction were now capable of facing the greater challenges of the tapering spire. It may be coincidental, but one of the very earliest experiments in spire construction was enacted above the Saxon tower in the leading quarrying village of Barnack.

The Decorated style, exemplified by one of the beautifully-carved capitals at Patrington church, Humberside.

The development of Gothic architecture continued through the remaining centuries of the Middle Ages, and even beyond. Around 1250, the Early English style, which had preserved a certain air of monastic asceticism and purity even in its more ornate cathedral applications, began to yield to the more exuberant, even playful flourishes of the 'Decorated'. This new manner of building gave a free rein to the accumulated skill of the medieval mason. Capitals which had formerly been embellished with the more formal flourishings of stiff-leaf decoration were now conceived as virtual hanging baskets, blossoming forth in vigorous outbursts of foliate decoration. Tracery, too, joined in the fun. Continued improvements in constructional techniques allowed the window spaces to be further enlarged and the plain 'plate tracery', which had pierced the gaps between Early English lancets, was superseded by the trefoil and quatrefoil patterns of geometrical tracery, and then by the less common petal-like openings of flowing tracery.

As the greater churches became more ornate and elaborate, so the labour force was diversified to include substantial numbers of carpenters, glaziers, polishers, smiths, marblers and plumbers. But England was now an extremely wealthy country. It needed to be. Dr A. J. Taylor estimates that more than $1\frac{1}{2}$ million silver pennies were struck to pay building wages at Beaumaris Castle during less than six months of the year 1295. Subsequently the pace and volume of building works gradually began to fall off. Economic decline, a worsening climate and soil exhaustion were forming the prelude to the disastrous Great Pestilence, which arrived in England in

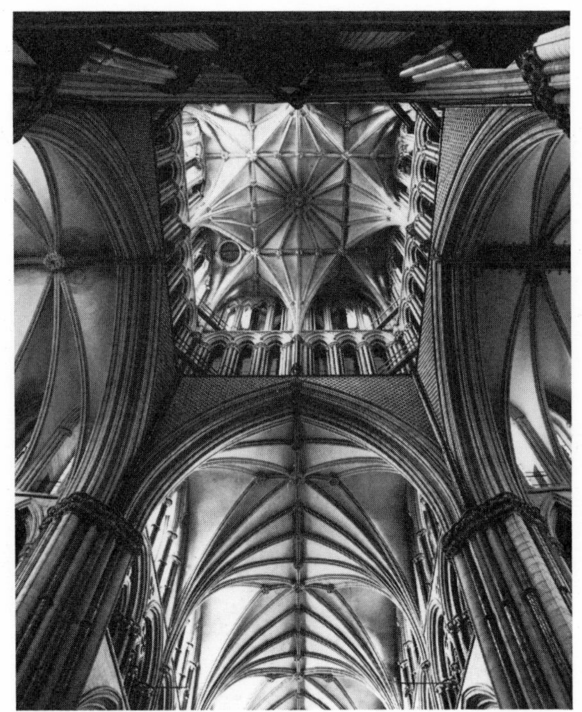

Magnificent vaulting at Lincoln Cathedral, accomplished, like so much of the best medieval architecture, in Jurassic limestone.

1348. Meanwhile, the humbler but fervent and extremely popular orders of Friars arriving in England from the Continent began to attract offerings and endowments which would otherwise have gone to the great established churches and abbeys.

The western half of England, which had been somewhat overshadowed by the east, now began to bear richer blossoms – expressed in magnificent cathedral-building works at Lichfield, Hereford, in the interior of Wells, but above all at Exeter. The architecture still mirrored the circumstances of the age: the continuing prosperity of the church, the fortunes which were beginning to accrue from the continental wool trade, the gradual emergence of a wealthy mercantile class, and, apparently, the desire of some churchmen to avoid taxes – which might be levied to fund wars – by investing their capital in the immovable assets of church masonry. In an age of increasing affluence, and one in which the church still exercised a powerful influence upon the minds of men both humble and powerful, it is not surprising that local landowning magnates were willing to underwrite immensely costly programmes of church building and, more particularly, church rebuilding and extension. Although the phase of greatest activity had taken place in the century before 1250, the abundance of Decorated features in English

The Pointed style, displayed in the ruins of East Creake Abbey in Norfolk.

parish churches testifies to the enduring urge to improve and extend. Much of the wealth which was petrified in the new works derived from the wool trade, and it cannot be entirely accidental that some of the most imposing parish churches of the Decorated period are associated with Midland and western sheep-rearing counties. However, they are much more closely linked with the numerous quarries of the Jurassic limestone belt, and underline the point that the costs of transporting stone remained a crucial and often prohibitive factor. Within the stone-rich areas, the local notables seem to have competed to provide the loftiest broach – and then parapet – spires, the liveliest tracery and the finest carving.

Let us not forget, however, that church masonry expresses but one facet of the social scene. It endures, while other more depressing aspects have perished. The centuries of wool-based prosperity were also the centuries when thousands of English villages were levelled and depopulated by flock-minded aristocrats. The churches may perpetuate the wealth of the wool trade, but only the merest bumps and hollows endure to mark the homes of villagers evicted to make way for sheep. The masons who built the churches were not rewarded according to their skills and their contributions to the making of the landscape – most remained both humble and vulnerable. Rather than preserving a truly Christian spirit, the lavish cathedrals, monasteries and churches tell of a religion that was institutionalized, and more concerned with assets, accountancy and splendid display than with the starvation and grinding deprivation that were the lot of the peasant. The Middle Ages were a violent, gloomy, heartless and doom-laden period, but the humorous and often satirical details which blossom on the carved wooden fittings, stone capitals and friezes of so many churches tell us that the carpenter and mason were not completely cowed by the circumstances of their age.

In 1348, a realm which now laboured under economic difficulties and was experiencing the ill effects of a steadily deteriorating climate confronted an infinitely greater horror: the first and greatest epidemic of the Pestilence. Already there were signs that the Decorated style was yielding to a new architectural form – which we know as the 'Perpendicular'. Deriving its impact from vast windows, painted glass, carpentry and the surging tower, rather than from the ornate details of its stone embellishment, the new style was in accord with an era of rocketing wage inflation and the thinning of the ranks of craftsmen which came in the aftermath of the Pestilence. While it is not difficult to relate the developing phases of the masonry tradition to changing social causes, it is less easy to explain the English adoption of the Perpendicular style. Previously, the developing nation had shared in continental building fashions and movements. Although the Perpendicular may have appeared in France before its application in Westminster, Old St Paul's and Gloucester Cathedral, it became the first truly national English style and was adopted and refined when the continental Decorated style evolved towards the florid extravagances of the 'Flamboyant'. Humbler manifestations of the Flamboyant style can be seen in some later medieval Irish churches, but in England, the Perpendicular became universal, surviving in the provinces and backwaters through several changes in fashion, and never completely dying.

A tendency to display stereotyped reproductions and a reduction in the quantity of carved stone – if not carved timber embellishments – for reasons of economy has meant that the Perpendicular style is sometimes derided as 'Businessman's Gothic'. This epithet was true in another sense, because the style evolved in an age when new wealth was seeking its expression in the landscape. It was an age of a rising capitalism, when an affluent middle class of merchants and entrepreneurs was beginning to make its presence apparent. In earlier centuries, members of the old aristocracy had sought to buy their tickets to heaven with endowments to cathedrals and monasteries. Considering the performances of the nobles when on earth, such passports would not have come cheaply! However, as the years rolled by, the wealthier members of society came to seek more personal expressions of their benevolence than the relative anonymity of a land grant to a

A number of the larger medieval stone buildings were found to be unstable, and at Wells Cathedral these reinforcing arches were needed to support the structure, which was in danger of collapse.

monastery or a bequest to a cathedral-building fund. Meanwhile, the increasingly immoral and avaricious conduct of many monastic communities undermined their credibility and appeal. Increasingly, the new wealth was invested in specific parish church and private chapel-building enterprises.

Both the monasteries and the cathedrals still enjoyed their independent sources of revenue. Although no completely new cathedrals were constructed in the Perpendicular period and style, it was used extensively from about 1350 onwards in reconstructions, and the construction of additions, vaulting and towers. A tower in the soaring new style was clearly much to be desired, and those at Durham – where the old tower had been conveniently ignited by lightning – Gloucester, Canterbury, Worcester and Wells are just as splendid as can be. Masons and stone carvers who may have regretted the passing of the intricate embellishments of the Decorated style, and who found the vertical mullions of Perpendicular windows less challenging and imaginative than the sweeps and points of flowing tracery, will at least have found stimulation and fulfilment in the construction of the cascading ribwork of the new roof vaults of cathedrals like Peterborough, Canterbury,

Wells Cathedral is noted for the magnificence of its Decorated architecture and carving.

Lavenham church is one of the most celebrated triumphs of the Perpendicular style of architecture.

Many of the wool towns and market centres of Wessex boast a superb parish church in the Perpendicular style; this is Steeple Aston in Wiltshire.

Norwich and Winchester.

There was also much work to be had in the host of village churches, where the benevolence or new status of the local landowning or mercantile dynasties were being dramatically proclaimed. In the rich, wool-producing and stone-rich lands of the Mendips, Cotswolds and Northampton Uplands, the great towers erupted, the side chapels swelled forth like golden buds – and the arms of noble benefactors were everywhere boldly displayed. The woollen industry was most heavily concentrated in East Anglia, but the best stones lay far away from here, in the Jurassic uplands of Lincolnshire and Northamptonshire, and even the richest benefactors could only afford to import enough for quoins and facings. At Lavenham church in Suffolk, which celebrates the marriage of the old wealth of the Earls of Oxford to the wool-based fortunes of the clothier Spring and Branch families, magnifi-

The Perpendicular church at Long Melford in Suffolk is noted for the magnificence of its flint and freestone flushwork.

cence was achieved in the form of walls of 'flushwork panelling', with squares of knapped local flint set within a framework of imported stone. The golden fleece of East Anglia materializes in the forms of many other great churches, like the flushwork palace at Long Melford – studded with memorials to the Clopton family – and the fine church at Stoke-by-Nayland, where the aristocratic inheritances of the landowning Tendrings and Howards merged with the newer wealth of local cloth merchants.

Despite the grandeur and perfection of the mason's craft which they enshrine, these Perpendicular churches belong to the approaching autumn of the stone-building era. In some ways, the later medieval masons had done their work too well. Virtually all but the small and remote communities of the northern uplands were served by a parish church or chapel. Most of these buildings were sufficiently improved and extended to accommodate the congregations of a slowly growing population. Most, too, were well-built; few needed major repairs before the eighteenth century – and few got them before the nineteenth.

Chapter 12

The Masons of the Middle Ages

While the importance of quarries to the development of the British landscape may be neglected, forgotten or taken for granted, the medieval mason has secured a niche in the popular imagination. Perhaps it is the vigorous survival of the organizations of Freemasons which have led to the perception of the medieval mason as a highly respected and influential craftsman who was initiated in almost mystical skills and covert practices, and cocooned by the corpus of patronage, restrictive standards and safeguards of the guild. While one cannot doubt that some of the awe surrounding medieval churches and cathedrals devolved on the men who constructed them, the typical mason was, in fact, not particularly well-paid or privileged, and was considerably less secure than the members of many other crafts. The misconceptions which surround the typical medieval mason reflect the lack of comprehensive information about individuals. Even the renowned master masons of the period remain shadowy figures, who can be identified at certain operations at particular times before receding into the mists of time. Since it is impossible to present a life history of a real medieval mason, the fictional character presented here is a composite of the influences and background of his day.

The story is set in the High Middle Ages, in the summer of the stone-building era, before the onslaught of the Pestilence.

Like many medieval people, Henry Easton takes his name from his native village, in this case a hilltop settlement amongst the freestone quarries of Northamptonshire. The names of masons often occur in building accounts, and many of these men have surnames which derive from quarrying districts. Knoop and Jones note that the building work at Westminster in 1292 involved the masons Edmunde Corfe, John de Corfe, Hugo de Corfe and Peter Corf – all taking their names from the stone-working settlement at the heart of the Isle of Purbeck industry.

Henry is conscious and grateful of the fact that he was born into a family of free peasants. His free status guarantees him neither employment nor freedom from starvation; like the other villagers, his kinsfolk live in poverty and in dread of famine and disease. Yet while they are expected to render labour and services to their lord, they do not live in bondage; they have access to the royal courts and enjoy a certain freedom of movement. Freedom is their most cherished asset.

Like other peasant children, Henry has had no schooling and is put to work at the earliest possible age. From time to time, his father leaves his holding of thirty scattered acre strips to earn welcome pennies as a labourer in one or other of the nearby quarries. As little more than a toddler, Henry accompanies him, performing little errands, carrying wedges, taking picks to the smith or gathering fuel for the lime-burners. While the quarrymen take their midday snooze, he wanders out of earshot and struggles to wield the hammer and pick; and all the time his absorbent young mind is saturated with the talk and actions around him. Most years, in the blustery months which follow Michaelmas, his uncle will arrive, footsore and unannounced, having walked ten, fifty or more than a hundred miles from some great building project to pass a few days of rest with the family. In the evening, in the dark protective gloom of the cramped homestead, the Eastons squat around the open hearth and, as the smoke streams swirl slowly above their heads, the storytelling begins.

The uncle tells of the great men he has seen, of bishops and abbots, of barons and their feuds and wars; of masters of works who interfere and snoop, and of the master masons, splendid in their livery, who can conceive a building in their minds, marshal the forces to manifest their idea in stone, timber and tiles, negotiate with paymasters, and forever monitor the output of each member of a small army of masons. Sitting silently entranced, Henry hears of the latest stages of work at the church project which has occupied his uncle for the last five years. He absorbs every detail. He learns of the people of the north and their strange, Danish-laden dialects, of the migrant armies of tough ditch diggers, of carters and carpenters, and of the

Modern masons engaged on restoration work still use tools and techniques which are very similar to those employed by the medieval masons.

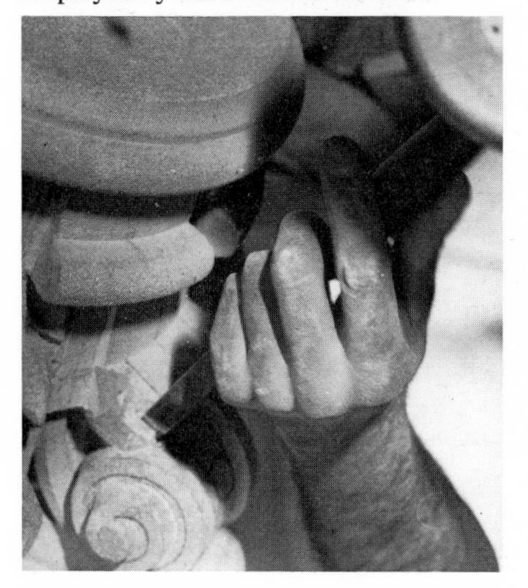

monks who look on. As soon as he is old enough, early in his teens, Henry joins his uncle for the arduous trek to the building site.

Their village contributes a steady trickle of young recruits to the ranks of the masons, all of them boys who have learned the basic skills in the nearby quarries and most of whom have relatives in the building trade. As the son of a free peasant, Henry is free to leave, but other intending workers in stone who come from the bonded classes must beg to leave under the licence of their lord, or flee the village under cover of darkness, never daring to return.

At the building site which is Henry's destination, there is no universal system of grading and payment in operation, although the many different qualities of craftsmen are paid according to a wide range of scales. At some great undertakings almost twenty different rates of pay are in operation, while at others there may be only five, six or seven. Similarly, there is no standard practice in the provision of tools: at some projects, he is expected to arrive carrying hammer and chisels; at others, these tools are freely provided, along with the services of a smithy for sharpening and forging them.

Henry is not able to enter upon a formal apprenticeship because, as yet,

The building, improvement and alteration of castles provided employment for thousands of medieval masons. At Goodrich Castle near Ross on Wye, the old keep tower survived inside the later defences.

none exists – there are no references to apprentices in masonry before the 1380s. A wide diversity of skills are practised at the building site. The nobleman who owns this – and many other villages – has appointed his own clerk of works to supervise the project. The design of the building, the employment of craftsmen, and the execution of the work are the responsibility of a master mason who is accountable to the clerk of works; and he is entrusted to enforce a code of discipline upon the various classes of stone workers. A corps of skilled journeyman masons works at the shaping and setting of the stones, and some of them have the assistance of particular servants or helpers, who are gradually learning their craft. Meanwhile, the crudest of the tasks at the building site and the quarries which serve it – work such as the lifting of stones, the carrying of stones and mortar, and some of the roughest stone-splitting and scappling – is performed by poorly-paid and unskilled local recruits.

Following a discussion between his uncle, who is a capable and respected mason, and the master, in which a truthful account of the lad's experience and character is given, Henry is appointed to the lowest pay grade in the ranks of the masons. He is instructed to assist the experienced workers in all tasks involved in the cutting and laying of stones, and to learn every facet of the craft. We cannot be sure what word Henry would have used to describe his new occupation. At some stage in the fourteenth century, the English word 'freemason' was sometimes adopted to describe the more skilful workers – accomplished in the working of the freestones, those which could be carved or sawn to a smooth-faced, ashlar surface. (Roughmasons worked with the less tractable stones, but it is likely that the two trades and terms overlapped considerably.) Henry may have used the Norman French term *masoun* to explain his craft, but in the accounting documents where his name would be listed (but which he was unable to read) he might at first be described as his uncle's servant, or *famulus*, and later as a *cementarius* or *lathomus*, both words associated with stone. It seems that, rather than being narrowly specialized, masons might be expected to work on either the shaping or the setting of stone, or even in quarrying, as the demands of the project dictated. Thus, terms which occur in old accounts – such as the *cissores* or *taylatores*, who were engaged in cutting stone, the *batarii*, who hammered or scappled stones, and the *cubitores* or *positores*, who layed or set the stones – might not reflect rigid divisions within the craft. The earlier Latin equivalent of the freestone masons, or freemasons, would be the *sculptores lapidum liberorum*, a term in use by the thirteenth century.

Just as the system of grading masons varied considerably from one building site to another, so too did the rates of pay. In 1351, the Statute of Labourers attempted, vainly, to set the rate at fourpence a day for the master mason, and threepence a day for other masons – rates which are thought

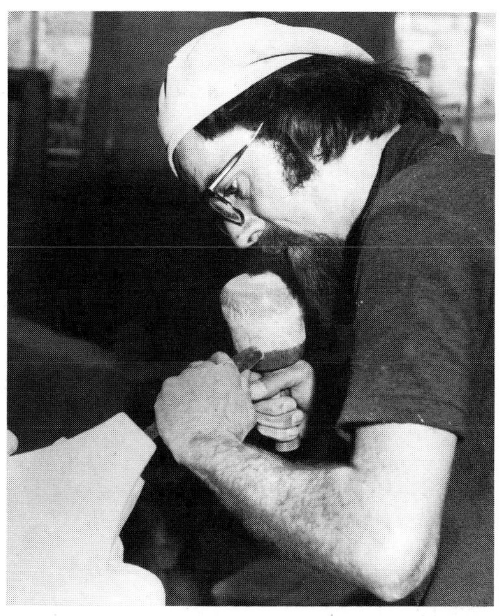

A modern mason working in the workshops of Rattee and Kett, Cambridge.

to have been normal before the Pestilence reduced the labour force and launched an era of shortage and wage inflation. In the thirteenth century, the average mason seems to have been paid at a daily rate of between twopence and fourpence, which did not normally include any provision of food. Those who have perceived the mason as a prestigious and well-paid guild member should note the rates quoted by Salzman from an early attempt in 1212 to regulate the wages in the London building trade. Freemasons, who represented the cream of their profession, were to be offered between twopence-ha'penny and fourpence per day, which was comparable to the rates of threepence with food or fourpence without food given to carpenters. Other masons and tilers received a similar rate, and their assistants earned from a penny-ha'penny to threepence. Plasterers received a penny less, while the less skilful ditch diggers and barrow pushers earned a rate which compares with that of the masons' assistants.

Various different sources suggest that, in the second half of the thirteenth century, craftsmen seem to have earned about threepence per day, labourers a penny less. A century later, the rates were about five pence and threepence-ha'penny respectively, in the fifteenth century, craftsmen earned about sixpence and labourers about fourpence a day.

In return for wages which were neither bountiful nor set at starvation levels, the mason was expected to work a twelve-hour day in summer and a five-and-a-half- or six-day week. In winter, the working day was reduced to nine hours, but often with an equivalent fall in wages. Only two brief breaks for eating or sleeping interrupted the working day, though some compensation was derived from the fact that up to twenty feast and holy days punctuated the year, and additional half-day holidays were sometimes allowed on the eves of the more important holy days. Occasionally, as when it was considered essential to complete a phase of work by a given date or when intense heat made exertions uncomfortable, incentive payments might be offered.

Henry begins his career on a wage of twopence a day, of which at least half is spent on food. However, given the length of the working day and the sleeping accommodation which is available in the temporary hostel erected on site for the masons, his expenditure is low, and he is able to save ha'pennies towards clothing and the purchase of his own setting tools, with a little left over to give to his mother. As his knowledge of masonry increases, so his thoughts turn to the future. To graduate to the elite ranks of the master masons seems an impossible aspiration, but Henry hopes that he may eventually become a senior and respected member of his craft – even an under master – while perhaps supplementing his wages by dealing in stone.

In the event, his promotion to a higher rate of pay comes in an unexpected and unwelcome manner. One scorching July day, when out of doors the heat of the sun rebounds from the glaring freestone and inside the masons' workshop or lodge the hot air is charged with stone dust which dries the throats and rasps the lungs, the workers earn a half-day holiday and an unexpected bonus. Several, Henry's uncle included, decide to celebrate in a neighbouring town. Off guard, their senses dulled by cheap ale, they are easily cornered in an alehouse by the sheriff's men and impressed for service at the site of a prospective royal castle. There is no right of appeal, no help which their master or employer can give, no point even in complaining. Henry assumes his uncle's work.

The vulnerability of the masons to impressment for royal service must have been one of the reasons for the retarded development of their guild system. (The first guilds of masons do not seem to have appeared until the second part of the fourteenth century, and the guild system was not firmly established in masonry before the medieval period had run its course.) The conscription of large numbers of these craftsmen for the construction of royal castles in Wales and elsewhere took place on an arbitrary basis. Many private and ecclesiastical undertakings must have been temporarily halted by the conscription of the workforces, and only the most powerful and influential of patrons could hope to recover masons thus impressed. Castle works sucked in labour from far afield. A famous passage from the *Anglo Saxon Chronicle* complains how, in the anarchy of Stephen's reign, 'They filled the land full of castles. They grievously oppressed the wretched men

This magnificent tower at Canterbury Cathedral recalls an age when architecture was the preserve of the master mason rather than the architect.

of the land with castle works.' The demands of the defence industry were to continue, and in 1277, for example, work on Aberystwyth castle drew in 120 masons, some of them from Dorset and Somerset. No guild system could be expected to exercise much power so long as its members were subject to conscription, or could be merged with forces of non-member conscripts. Presumably, the sheriffs were able to retain the masons they had conscripted, for few seem to have escaped; perhaps the threat of imprisonment was a sufficient deterrent. However, many masons will have considered themselves just as well employed at the royal works as at other sites. It must be remembered that the guild was not a national organization comparable to a modern trade union. Each town had its own craft and trade guilds, and membership was not interchangeable.

This, in fact, is probably the most important reason for the late development of guilds in masonry. While the guilds were essentially town-based, masons frequently worked on large-scale royal or ecclesiastical projects which lay some miles from the nearest town. Stone buildings other than churches were rare in the towns of the Middle Ages; the mason could not set up his workshop in a town and expect a steady stream of orders – as could the locksmith or housewright. Instead, he was obliged to travel to where his skills would be welcomed and was thus essentially a journeyman, who might travel a hundred miles to a site. For the greater part of the Middle Ages, the guild was substituted by the masons' lodge, a workshop which was erected at the site of an important building operation, with accommodation for one to two dozen craftsmen. Sometimes, the lodge seems to have provided sleeping as well as working accommodation, but in other cases the masons lived in separate hostels. The master mason and his deputies were responsible for maintaining discipline within the lodge, and the master had the power to fine those members of the lodge who misbehaved. There can be no doubt that most masons were accomplished and hard-working craftsmen; but although medieval society was repressive, ordered and hierarchical, rowdiness, indiscipline and brawling were also common. Some lodges appear to have operated with a minimum of friction, but at others, such as that established for the building of Eton College, petty squabbles seem to have been frequent.

The ultimate sanction which the master mason could impose was dismissal. If wise, he will have taken account of the difficulty of finding replacements, for there must have been many periods during the cathedral- and castle-building eras and after the Great Pestilence of 1348–9 when skilled masons were in short supply. Since the typical mason was accustomed to walking many miles to a place of employment, it is likely that he will have neither welcomed dismissal, nor considered it likely to damage his future prospects – a bad reputation could be left behind. Just as it would be wrong to assume that all masons were virtuous and all lodges well organized, it would be wrong to accept that all master masons, setters and cutters were skilled and infallible: plenty of buildings fell down or needed extensive repairs as a result of bad design or workmanship.

As a typical member of his calling, Henry Easton wisely accepted his niche in the hierarchy of the lodge, attempted to assimilate the skills which he witnessed, and took great pride and satisfaction from the respect he won from his peers. As his experience and craftsmanship accrued, so he progressed from rough work with the scappling hammer to tooling stones with the boaster, and was eventually entrusted with the carving of capitals and mouldings with the chisel. Even so, Henry may never have had an individual mason's mark. As Salzman points out, much, including some fanciful nonsense, has been written about masons' marks. The stones in many medieval buildings bear the marks of the masons who worked them, but rather than being the mystical symbols of high craftsmanship, they may only be the markings of less illustrious masons who were employed on piecework. In this case, they may have been engraved to assist the paymaster in his calculations, but whether the marked stones were shaped in the quarry or in the masons' lodge is not known. (Wells Cathedral has a proliferation of masons' marks and a printed guide is on sale there.)

In the course of his career, Henry works on numerous, widely-scattered projects, serving for weeks, months or years according to the nature of the work. From time to time, his plans and ambitions must be swept aside as a result of his conscription for work on the king's buildings. In the depths of winter, when masons are laid off because of adverse conditions of light and weather, or between the completion of one task and the start of another, Henry makes the long journey to his native village, trudging through the rain and snow, hitching lifts on stone carts, sometimes sleeping in barns or hedgerows. Following one such expedition, he returns to find that his father is dead and buried and his mother is struggling to work the family holding. He postpones his departure until the open strips are ploughed, and chooses a bride from a nearby village – a robust and capable girl who can perform the farmwork during his long absences from the tenancy which is now his own.

Thereafter, Henry always attempts to spend a few winter months at home. His work as a mason alternates with weeks of ploughing and sowing, threshing, hedging and ditching. Although Henry is a skilled and respected mason and his wages provide a small buttress against famine – allowing the purchase of seed in seasons of shortage and a milk cow to feed the young children – the family's standard of living seldom rises much above that of his neighbours. By the time he is fifty, Henry's working life is almost over. Long exposure to rain, sleet and frost has taken its toll – arthritis has stiffened his joints, and the stone dust which clouds the atmosphere of the lodge has ravaged his lungs. Retirement is a luxury which Henry cannot afford, but after a few seasons of light farmwork on his holding, like so many of his former workmates, he dies of silicosis.

Like the first he can remember, Henry's last evenings are spent in yarning around the open hearth. He is now the storyteller, and his breathless, rasping

Open air workshops, such as this one at Lincoln, were a common sight at medieval building operations.

voice gains strength as he tells of glittering cathedrals and cliff-faced ramparts, distant shires and former companions. His second son inherits the chisels and squares before retracing his father's steps from the nearby quarries to the far-off building sites.

In common with all other medieval masons, apart from a small élite amongst the masters, Henry knew neither fame nor even moderate wealth. But despite his relative poverty and anonymity, Henry lived a life which was infinitely more colourful and varied than those lived by his contemporaries and equals who remained village peasants. Though buried in a shallow, unmarked grave in the Easton churchyard, he and his workmates could share the epitaph to Sir Christopher Wren which is inscribed in St Paul's Cathedral: 'If you would seek his monument look around.'

Most masons will have taken enormous pride in their individual and collective achievements as the creators of the most ornate and spell-binding constructions Europe has ever seen. Specific memorials to this reverence for craftsmanship are hard to find, but scratched on a pillar by the south door of Ashwell church in Hertfordshire is the Latin graffito: '*Cornua non sunt arto compungente – sputuo.*' These may well be the words of a visiting master mason who was affronted by the quality of workmanship on display: 'These corners are not straight – I spit at them.'

Chapter 13

From Autumn into Winter

We have seen that the stone buildings of medieval England were manifestations of the beliefs and social currents of an evolving nation. From the point of view of the craft of masonry, the medieval climate could hardly have been more salubrious. The gross social inequalities and injustices which concentrated the nation's wealth in the coffers of the church, the monarchy and a small minority of aristocrats thus provided the wherewithal for gigantic building programmes. The results were 'establishment' buildings – cathedrals, monasteries, royal and baronial citadels – petrified expressions of the values of a tiny élite among men, who controlled great resources and power and financed the sorts of schemes which they considered good for themselves and good for the country. If the medieval environment for great building works was ideal, then any social changes that followed could only have an adverse effect on the British stone building tradition.

The traumas which signalled the close of the Middle Ages encouraged some new enterprises in masonry, but removed more incentives than they created. The transformations affecting the status of the church were crucial. Although the pace of monastic building was declining in the later medieval centuries, new works were being commissioned right up to the suppression of the smaller houses in 1536, and of the larger monasteries in 1539. It is true that the stone-built country mansion was beginning to make its presence felt in the English landscape at this time; but nevertheless, a vital source of employment for masons in ecclesiastical building had been lost at a stroke with the Dissolution of the Monasteries. What is more, the abundance of dressed stones which were there almost for the taking at hundreds of monastic houses must have undermined trade for many quarrymen and stone dressers. Monasticism, however, was already in decline. As religious foundations had accumulated vast estates and remarkable revenues, their early spiritual intensity, which had drawn many recruits to the monastic lifestyle, had been diluted. At the same time, repeated outbreaks of the Pestilence had ravaged religious communities no less than towns and villages. Most of the larger monasteries were half-empty. Many of the monks' practices were now ridiculed rather than revered. Nevertheless the movement had sufficient vitality and enough sympathisers for the restoration of the suppressed abbeys to be one of the goals of the northern rebellion of 1536, known as the Pilgrimage of Grace, and of smaller uprisings in the south-west.

At Fountains Abbey in Yorkshire, the changes are reflected in the history of the stones. This was the richest of the Cistercian monasteries in Britain, with an annual revenue of over £1,000. In 1539, the Abbey was surrendered into the hands of the king, and its community was pensioned off with sums ranging from Abbot Bradley's £100 a year to an income of £5 each for the lowliest of the monks. The Fountains' estates were swiftly sold to Sir Richard Gresham and, in 1597, Stephen Proctor, son of a successful ironmaster and a member of the rising class of *nouveaux riches* entrepreneurs, bought the abbey from the Greshams. New wealth demanded the gentlemanly status associated with a country mansion. Within a few years, Proctor had begun robbing the abbey ruins of its whitish freestone building blocks and darker embellishments of 'Nidderdale marble' (a fossiliferous limestone which was the local equivalent of Purbeck marble). With these pillaged materials, his masons built Fountains Hall, and it is by way of the solidly stately frontage of this mansion that the visitor approaches the abbey ruins.

No less typical of the changing social order was John Gostwick, who rose from the ranks of the yeoman class to become Master of Horse under Cardinal Wolsey and a Treasurer under Henry VIII, for which, like Proctor, he was duly knighted. He was involved in the dissolution of monastic houses in the area of Bedford, and it is not hard to imagine what happened to their stones. Gostwick's manor has gone, but its impressive outbuildings – a fine stables and exceptional dovecote – remain, along with the late Perpendicular church – or at the very least its chapel – which his tombstone tells he had built.

The cathedrals fared a little better than the monasteries, though with their images and splendid ornaments they too were often regarded as 'Popish symbols'. Neglect was the best that most could hope for, and many suffered at the hands of iconoclasts – a triumph of dogma over beauty that closely parallels the destruction of much of our heritage by Philistine public policies today. Iconoclasm firstly affected the monasteries, and then, in the reign of Elizabeth I, renewed outbreaks reflected the tensions between the Protestant state and the powerful Roman influences; widespread and systematic assaults on the heritage of images, crosses and carvings were then made

With materials plundered from Fountains Abbey (above) the entrepreneur Stephen Proctor built Fountains Hall (below).

John Gostwick's dovecote at Willington in Bedfordshire, built from materials plundered from neighbouring abbeys (a National Trust property).

during the Cromwell years. The smashing of painted glass and images in churches was matched by the breaking of stone crosses in villages and on highways and bridges. Almost every market, many river crossings and various boundaries and places of wayside incident sported a cross. The destruction of these symbols, which had served liberally as signposts, markers and insurance policies for securing divine protection, was as thorough as it seems now to have been mindless. At least a score of cathedrals were vandalized in one way or another during the disturbed years of the Reformation and Commonwealth.

Thereafter, the cathedrals tended to be ill-maintained. The age of cathedral building was all but over. Only one was built in the seventeenth and eighteenth centuries: St Paul's. The original cathedral was in an advanced state of decay when it was damaged in the Fire of London of 1666. It has its humble epitaph in stone – and it is also represented in a graffito scratched into the soft chalk clunch inside the tower of Ashwell village church in Hertfordshire.

Partly because of the relatively recent date of work on the new St Paul's Cathedral, much detailed documentation has survived, and has been subjected to scholarly analysis by J. H. Bettey. From the points of view of Wren, the architect, his Commissioner associates, and the quarrying communities of the island of Portland, the story is one of unremitting frustration. High grade Portland oolite had been used on important building works since at least the fourteenth century and the stone came into special prominence when it was adopted by Inigo Jones for various prestigious buildings, including part of the Banqueting Hall at Whitehall.

However, quarrying on the island was conditioned by the survival of firmly-rooted medieval customs. The 198 tenants of the royal manor on Portland were deeply distrustful of outsiders, or 'Kimberlins', while quarrying was subject to ancient customs and disputes were settled at a court leet (which has survived into modern times). Some quarries lay on demesne land but all tenants had the right to open quarries on the commons of the manor, subject only to the payment of dues for the stone taken. Almost throughout the quarrying operations, work was interrupted and frustrated by squabbles concerning the payment to tenants of duty for stone taken from the commons; disputes about the violation of ancient rights; arguments about the employment of kimberlins in the Portland quarries, and the extreme dislike by the islanders of the agents appointed by the Commissioners – their practices sometimes seeming less than honest.

The quarrying work could not be forced ahead in the face of all these squabbles since the law seemed clearly to favour the Portlanders' case. In the early months of 1678, the antagonism exploded in strikes and riots, with carts, impliments, cranes, piers and new roads being smashed up so that work at the cathedral ground to a halt. The four ringleaders were summoned to appear before the king at Whitehall, but were given what amounted to a conditional discharge.

There were other problems, too, of a serious and varied nature. The cathedral operation transformed the scale of quarrying at Portland and the need to maintain a steady supply of stone strained the shipping resources of the nation. In 1696, a serious landslide destroyed roads, cranes and jetties essential to the loading of the waiting ships, and the delivery of stone was affected for more than two years thereafter. In the years around the turn of the century, privateers and French raiders took a steady toll of the lumbering stone ships, while the ships themselves posed problems. Vessels of a suitably large size were in short supply, and some had to have their holds and hatchways altered to accommodate the larger sizes of stone. Also, there were problems of quality control, and although wooden patterns of the more specialised stones were distributed to the masons and quarrymen of Portland, on arrival in London many stones were found to be of inferior quality or inadequate.

Work was finally completed in 1714 – and all concerned must have sighed deeply in relief. Wren himself was clearly sickened by the interminable catalogue of problems, and in 1705 he had written to the quarrymen '. . . though 'tis in your power to be as ungrateful as you will, yet you must not think that your insolence will be always borne with. . .' These were the words of a man whose work had been frustrated by the accidents of history and who doubted the legal force of his own complaints.

The Portland documents are also interesting because they reveal contemporary wages and prices. Quarry workers were offered 20d per day. Small stones of less than three tons were sold at 11s 6d per ton for scappled stone and 10s 2d for ordinary block. These prices did not include transport and costs rose steeply as the sizes of stone increased, so that stones of 7 tons sold at 22s per ton.

After the Reformation, changes in belief were accompanied by changes in architectural fashion. It seems that the arbiters of taste had become blind, or bored, in matters concerning their heritage of splendid buildings. Medieval folk had gazed in wonder at the splendours of the mushrooming ecclesiastical architecture, as the description of the 'wonderly wel-y-bild' church in the *Crede of Piers Plowman* reminds us:

> With niches on everiche half,
> And bellyche y-corven;
> With crochetes on corneres,
> With knottes of gold,
> With gay glittering glas
> Glowying as the sunne . . .

As the waves of the Continental Renaissance lapped upon the English consciousness, and fashion romanticized the lost civilizations of Rome and Greece, the remarkable advances and achievements of the Middle Ages were often held as nought. John Evelyn was no common Philistine; indeed, in his early advocacy of tree-planting, he was almost a conservationist. Yet he summed up the fashionable seventeenth-century attitude to the medieval building legacy when he castigated it as '. . . a certain fantastical and licentious manner of building: congestions of heavy, dark, melancholy, monkish piles, without any just proportion, use or beauty.'

The names of the sub-divisions of medieval architecture – Transitional, Early English and so on – were coined by Thomas Rickman in 1817. The term 'Gothic' is a couple of centuries older; it was used as a term of derision for the medieval styles and to convey a judgement of barbarism – in the way that the word 'Hun' has ignorantly been used by the English against the Germans and by the Scots against the English.

In all the architectural changes, fashion and social and intellectual transformation were fearfully intertwined. The disciplined symmetry of the neo-Classical buildings symbolized both a sentimental infatuation with the lost Classical civilizations, and the desire to create an orderly and enlightened society out of the wreckage and traumas of the post-medieval world. The sundering of the power of Rome over thought and philosophy freed the more innovative members of society to experiment with the architecture and theology of pagan civilizations. The native building heritage was also the heritage of the restrictive and rejected Roman Church – thus it is hardly surprising that fashion-conscious men and women were blinded to the great churches' glories, and to the possibility of continuing to develop the great architectural tradition.

The age which gave birth to the neo-Classical architectural tradition was also the one in which the master mason began to yield to the architect in the design and planning of buildings. Is it not ironic that the incredibly elaborate and individual buildings which make up the medieval legacy were the creations of craftsmen, who generally lacked any formal training in the disciplines of art and science, while the plain, repetitive and strictly-proportioned buildings of the neo-Classical movement were increasingly the work of architects? Almost anyone endowed with a measure of patience can reproduce a Corinthian or Ionic column. Only a man or woman of great genius could conceive a cathedral to compare with the intricate marvels of Exeter, Wells or Lincoln.

From the point of view of the stone carver, the general suspension of church and cathedral building, and the plainness of the relatively few neo-Classical churches which were built, will have been disastrous. The employment for masons in general must have declined, although the growing vogue for palatial country mansions provided many new outlets. During the turbulent years of the High Middle Ages, the greater nobles tended to live in massive, draughty castles, behind such thicknesses of protective stone walling as they could afford. The lesser nobles and lords of manors lived, on the whole, in halls and manor houses of timber framing rather than of stone, often with an encircling moat to provide at least the trappings of status and

The loss of employment for masons on ecclesiastical building works was partly compensated for by the new vogue for oppulent country mansions. Castle Ashby in Northamptonshire was begun in 1574 and not completed until the following century was well advanced, so that it incorporated both Elizabethan and Jacobean styles.

the illusion of defence. As the medieval period wore on, the castle began to lose its defensive credibility. A number of factors were involved: amongst them were the irresistible size of the king's mercenary armies, the problems caused by the improving cannon, and the emergence from the chaos of the Wars of the Roses of a masterful Tudor dynasty, which was prepared to brook less nonsense from castle-brooding provincial potentates. With the castle slithering into obsolescence, the church less magnetic than before, but the estate revenues still rolling in, it is not surprising that so many later medieval nobles began to invest in the accompaniments of stylish and comfortable country living. The dying castle gave birth to the stately home.

Timber-framed halls and manor houses are certainly as beautiful as others built of stone, and the details of their craftsmanship match most of the achievements of the mason. But in practical terms, the timber-framed mansion had certain drawbacks for affluent local aristocrats and *nouveaux-riches* converts to country life who sought convenience and elegance. Although it could be lengthened *ad infinitum* by the addition of new bays at either end, and additions could be added at right angles to embrace a square courtyard or to give the house an L-, E- or double E-shaped plan, it was not easy to fashion a timber-framed dwelling which was more than one room in width – the width of the bays being, to a considerable extent, controlled by the lengths of available timbers. As the house grew longer and gained more right-angled additions, the problems of internal movement and the difficulties of providing corridors increased. When the new vogue for symmetrical houses in the Classical manner was established, the rambling, twisting timbered home, with its undisciplined proportions and none of its lines exactly vertical or horizontal, came to be considered *infra dig.* (Fortunately, some lords were sufficiently impecunious or impervious to fashion to prevent its extinction.)

Towards the end of the Middle Ages, mushrooming country mansions were providing employment for many of the masons released by the decline in ecclesiastical building. The stone mansion, however, was not without its rival. In the 1480s, in a realm which was recovering from the destruction of the Wars of the Roses, work began on one of the first and noblest of country mansions at Compton Wynyates in Warwickshire. The psychological association between status and defenceworks still exerted an influence, and so the mansion was provided with a moat. The Classical fashions were yet to infect English architecture, and therefore the splendid late-medieval house was unbridled by the constraints of symmetry. In the 1520s, Compton Wynyates was enlarged: lofty chimneys proclaimed the new comforts of modern living, and vast new windows announced that illumination was more important than defence. Compton Wynyates is notable in another way. It is built of brick.

In 1514, work began on the construction of a retreat for England's Cardinal, Thomas Wolsey, supreme in government and second only to the king in power. Renaissance symmetry was evident in the façade of his Hampton Court Palace, but the incorporation of many Gothic details enlivened the effect. The scale was much greater than that of Compton Wynyates, and the building work provided employment for 2,500 men for five years. Hampton Court too was built of brick.

Many years were to pass before brick would be regarded as a cheaper substitute for stone, but it *was* a fashionable and prestigious alternative, used at first only for 'castles' and great houses. Chimneys, too, were something of an innovation, and the fancy 'barley sugar' twists and projecting courses which were built into these Tudor status symbols testify to the bricklayers' rapid mastery of their craft. Since the techniques are not dissimilar, we can assume that many bricklayers were converts from the craft of stone laying. The choice between brick and stone came to depend simply on the foibles of those who footed the bills and the relative proximity of the building site to beds of freestone and brick clays.

In the traditional quarrying counties, stone held sway. But while beds of Oxford Clay were mined in Bedfordshire, Buckinghamshire and Hunting-

Towards the end of the medieval period, brick emerged as a prestigious rival to stone and many masons must have switched to the craft of bricklaying. Oxborough Hall in Norfolk was begun in 1482 and displays an early mastery of the craft (a National Trust property).

donshire, in Northamptonshire, they were there to be exploited but were largely ignored. In this county, the magnificent houses like Kirby Hall, Lilford Hall and Castle Ashby proclaimed the splendour of the native limestone. Counties like Buckinghamshire were in an intermediate position: although the greater part of the county is poor in stone, the Rivers Thames, Cherwell and Ouse gave access to the limestones of the Cotswolds, Oxfordshire, Northamptonshire, Rutland and Lincolnshire. In the fifteenth-, sixteenth- and seventeenth-century buildings of Eton College, one finds sixteenth-century brickwork combined with the stones of Taynton in Oxfordshire, those of the Greensand beds of Merstham in Surrey, and the fine white oolite of Caen in France. In counties like Suffolk, which had poorer access to building stone, any house-building operations employing the mightily expensive imported stones was likely to prove considerably more costly than if brick were used. Timber-framing for the smaller mansion, and second-rate local stones like flint or clunch offered the cheapest alternatives.

At Long Melford in Suffolk, we have seen how luxurious imported freestone and cheap local flints combined in the panelled flushwork of the remarkable fifteenth-century church. About 1560, much of the wealth amassed by the artful and successful lawyer-politician William Cordell was invested in the construction here of an imposing hall. The new hall was sufficiently lavish for Queen Elizabeth to be entertained in it in 1578, and, in an amazing display of his wealth, Cordell provided a retinue of some 2,000 men to greet her on her arrival. Yet in the construction of Long Melford Hall, as in the building of so many stately homes in the stone-poor counties, Cordell opted for brick; the enormous green in the village still undulates, revealing the hollow which marks the pit from which the brick clays were dug.

It was not only a selection of the stately mansions, palaces and the Tudor coastal bastions that were built of stone in the decades following the suspension of most ecclesiastical construction. Work continued apace in the provision of new or extended college buildings at the two English university towns of Oxford and Cambridge. Here again, the wonderful limestones of the Midlands oolite belt were widely used. They could be quarried almost on the doorstep of Oxford, while the networks of Fenland rivers and narrow canals, or 'lodes', facilitated the transport of stones from quarries like Barnack, Ketton and Weldon southwards to Cambridge. The new availability of cheap stones, which were robbed from the suppressed monastic houses, stimulated the college works. Thus at Cambridge, for example, Trinity College used the former Grey Friars convent as a quarry in the 1550s, and took stones from Ramsey Abbey in the 1560s; King's College also quarried Ramsey for stones used in rebuilding the college hall; while in 1579, Corpus Christi chapel was constructed of stones from Thorney Abbey and Barnwell Priory.

So far in this brief history, we have been preoccupied with great ecclesiastical buildings and the castles and mansions of men of wealth and power. Until the close of the Middle Ages, the humble stone dwelling was almost everywhere rare and unusual. A number of medieval towns were walled in stone rubble: the Roman and medieval walls of York, built with a brilliant white Magnesian Limestone and restored in the nineteenth century, provide the best and most neatly-dressed example. Many town walls must have been built rather to proclaim the status of the town than to secure against attack, but the same forces which rendered the castle redundant also devalued the town wall which had become a cramping corset on development. Where such walls existed, the houses, workshops and warehouses within were generally of timber, wattle and thatch. In the Tudor centuries and those which followed, towns expanded beyond the restrictive bondage of their walls, and the townsfolk were sometimes wont to pillage the defenceworks for stone. As well as robbing abbey and convent ruins, King's College seems to have used stones from the great hall of Cambridge Castle, while Knoop and Jones record cases involving the theft of stones from the town wall of Leicester, and from London City wall.

Building work at the great university centres of Oxford and Cambridge continued apace from the late medieval centuries through into the post-Reformation era. Consequently both Gothic and Classical architecture of the highest quality can be explored. The Classical styles, however, offered fewer opportunities for decorative stonework. ABOVE Intricate Tudor decoration at King's College chapel, Cambridge. BELOW The less ornate neo-classical masonry of the Old Clarendon Building, Oxford, dating from 1711–13.

In the late sixteenth century, William Harrison wrote: 'The greater part of our building in the cities and good towns of England consisteth onelie of timber, for as yet few of the houses of the communaltie (except here and there in the Welsh countrie townes) are made of stone.' He added that 'The ancient manors and houses of our gentlemen are yet, and for the most part of strong timber,' but went on to mention that 'Howbeit such as be latelie builded are commonlie either of bricks or hard stone.' Harrison was, in fact, noting the early stages of a major social movement which W. G. Hoskins has christened 'The Great Rebuilding'.

There were other pressing reasons why the traditional town of thatched and timbered dwellings and filthy rutted streets should be improved. As early as the fifteenth century, the streets of Canterbury, Exeter, Gloucester and Southampton were partly paved in stone so as to be cleaner and more easily passable. More compelling was the hazard of fire. In the course of the seven decades which preceded the Great Fire of 1666, various efforts were made to encourage the use of less inflammable buildings in London, and after the conflagration a statute required that the external walls of all new buildings should be of brick or stone. In the provinces, the towns where sixteenth- and seventeenth-century domestic building is extensively preserved tend to

Changing tastes and circumstances are epitomised by this inn at Montgomery. It was built of stone plundered from the ruins of the medieval castle and existed as a stone building for several centuries, until the 'trendy' mock timber-framing was added as a façade in quite recent times.

be closely linked to the traditional quarrying areas: Stamford, where the walls of Barnack stone are graced with roofs of Collyweston slate, and Oundle in Northamptonshire are superb, while the former glories of little Bruton in Somerset are evident in a fine legacy of buildings which display the wares of the famous Ham Hill quarries.

A yawning social chasm separated the late-medieval burgher from the feudal peasant. Throughout the Middle Ages, the peasant lived in a cramped hovel of one or two rooms: the family and its few livestock often shared a single roof, and the house could be so poorly constructed that it might need rebuilding every generation. Even in the quarrying districts, these dwellings were almost invariably built of mud, sticks, wattle, daub and thatch, though in some areas, where rubble of a soft stone like chalk or a scatter of hard granite or gritstone boulders could be respectively hacked or gathered from the site, the home might stand on rubble footings. Such pitiful hovels were not built to endure, and so our best knowledge of them has been won from the excavation of lost village sites like Wharram Percy in Yorkshire.

As the rigidities of feudalism yielded to the greater fluidity of the emergent capitalist society, so opportunities for self-betterment could be exploited by individuals from the more affluent or enterprising sectors of peasant society. Free peasants, labourers, artisans and yeomen, whose humble lives had previously been battles for survival, now began to aspire to more homely comforts. It was against this background of modestly rising expectations and affluence that the Great Rebuilding took place. In the clay and chalk lands, many cramped and makeshift cruck- or box-framed farmsteads were replaced by larger, sturdier box-framed dwellings which were built to last. So well were these constructed that hundreds still grace the lowland landscape. In stony uplands and some quarrying districts, the change from a home of timber to a new one of stone was often made.

The Great Rebuilding originated in the south-east of England about the middle of the sixteenth century, with the waves of change lapping slowly across the counties to reach the northern and western provinces a century or more later. Even in stone-rich backwaters like the gritstone country of the Pennine Dales, the new generations of farmsteads and cottages were not always built of stone, and villages of cruck-framed and thatched dwellings could still be seen in places like Nidderdale at the start of the nineteenth century. Cornwall only became a county of stone farmsteads after the Middle Ages were over and gone.

It would be folly to try to rank the merits of the vernacular traditions of dwellings in timber-framing, brick or stone. The timber tradition has its own enticing regional variations of design, while old brickwork can have a wonderful richness of texture and hue. The great appeal of the humbler

Gladstones' Land in Edinburgh was rebuilt in 1617 by the merchant Thomas Gladstones as a tenement of middle-class apartments, and is one of the oldest stone-built domestic buildings surviving in Scotland.

stone-built dwellings is that they fit in so harmoniously with the embracing landscape. Seldom did their owners court a fashionable or a picturesque appearance. When they did, bleak miniatures of the Classical house or florid parodies of vernacular styles were often the result. Many of the attractions of seventeenth-, eighteenth- and nineteenth-century cottages result from the limitations on their building budgets. The importation of prestigious stones proving far too costly, they are built from the good or second-rate stones which were close at hand. As a result, they match and mirror the local environment. Thus, in parts of Cornwall, the granite farmsteads are made of moorstone, gathered and assembled into houses which echo the steely hues and textures of the boulder-strewn moors, rugged field walls and knobbly granite tors of the skyline. In the more mellow farmlands of the Cotswolds and Northamptonshire, the rich, honey tones of the buildings are repeated in drystone field walls and ploughlands; in the chalklands, flints

In the eighteenth century, Bath began to emerge as a stylish stone-built town. Supplies of top class (if rather perishable) building stone were available at nearby quarries, which also exported their wares far and wide.

flash in field scatters, and also in cottages and church walls; while in Scotland, Ireland, Wales and the northern uplands, the tones and textures of the scree slope, crag, scar, river beach and stream bed outcrop again in the walls of fields and farms.

While the great quarries which punctuated the oolite belt exported their products far and wide, home-making villagers often exploited the less prestigious local ragstones and ironstones of the intervening spaces. Thus, every other village tends to have its own particular subtlety of tone – from the gingerbread of the ironstone localities, through the greys and buffs of rag-stone, to the milky oolites and fossil-spangled 'marbles'. On the margins of so many of these places, there are the rumble-tumble undulations which mark the roughly filled-in pits where the village stone was quarried. Frugal use of the funds for house-building also ensured that the majority of stone cottages and farmsteads were built by local roughmasons from coursed rubble, rather than by freestone masons who might furnish smooth-faced homes of squared ashlar blocks. In so many villages of the stone counties, one may see old patterns of affluence and economy preserved in the landscape, with the slightly larger eighteenth-century cottages, built in ashlar and with a little bit of fancy work in the pediment above the door, sitting cheek-by-jowl with the smaller, cheaper dwellings in coursed rubble. On the whole, it is the latter class of dwellings, with their rougher textures, which sit most snugly in the rural scene.

Each stone-blessed locality in Britain developed its own distinctive vernacular building tradition. Let us look at just one such tradition – that of the Millstone Grit area of the Yorkshire Dales. The ancient origins of the stone-built peasant dwelling are evident in the tumbled ruins of circular Iron Age huts which can be found in the limestone country around Grassington and Malham. These huts play a dubious part in the lineage of circular, rubble-walled and thatch-roofed shepherds' huts which once existed in the Dales (according to sketches drawn on late-medieval manuscripts), but they did not influence later cottage designs. During the medieval period, a number of the greater buildings were built of stone: the abbeys and castles, of course, whose occupants between them owned most of the land, but also a few lesser fortified houses, like Nappa Hall in Wensleydale, and some of the halls of the gentry, like Markenfield Hall near Ripon. However, the smaller farmsteads and cottages of the Dales seem to result from the translation into stone of vernacular building designs which were originally accomplished in timber-framing, wattle and thatch.

The most influential traditional house form in the Dales was that of the long-house. Some writers would trace its pedigree straight back to the long-house homesteads of Dark Age Norse settlers. However, the evidence from the deserted medieval villages shows that the long-house (in which the oblong space beneath a long narrow roof was partitioned to provide family quarters and a cow byre) was also a very common form of peasant dwelling in areas unaffected by Norse settlement. In its medieval form, the long-house would generally have been a cramped hovel, about the size of a small bus and permanently pervaded by the stench from its byre compartment – which, for obvious reasons of hygiene, was sited at the downslope end of the dwelling. As the long-house evolved, it became more spacious and the barn and living areas were more firmly partitioned, each being served by

Many farmsteads survive to testify to the arrival of the Great Rebuilding in the Yorkshire Dales. The main section of this one in Coverdale dates from the early eighteenth century and appears to have been added to an older building.

a separate access from outside. A typical sixteenth-century example might be around seventy feet in length, though only about twenty feet wide. The downslope end of the farmstead would still be occupied by a compartment with cattle stalls, which is known in different Dales localities as a 'shippon', 'byre', 'mistal' or 'boise'. Next might come a barn with doors at either side, which could be opened during threshing on windy days to assist the winnowing of grain. Finally, at the upslope end of the dwelling, was the living room, sometimes partitioned to provide a separate parlour, and perhaps with a projecting kitchen or pantry addition.

Like the rectangular hay barns standing isolated in the meadows, the dwellings of the medieval peasant and yeoman were built in the time-honoured cruck-framing manner. The essential structure consisted of a pair of inward-arching oaken cruck blades, which were set at either end of the building to form the inverted V-shaped supports which carried the beam of the ridge tree. Because the cruck blades curved inwards to form the gables of the roof projecting horizontal tie beams were fixed to the crucks at the height of the intended eaves to give a width at this level equal to that at the base of the cruck. These tie beams then provided supports to carry the long beams or pans which anchored the bases of the rafters, and the lesser timbers which carried the mud-coated lattice of branches known as wattle and daub.

Even in the medieval period, however, the footings and lower levels of the walls may commonly have been of a rubble of stones gathered from

Vernacular traditions here are represented by the use of local grit and flagstones for walling and roofing and the projecting 'kneelers' at the bottom of the gables. This dwelling in Nidderdale was probably built around 1800 as a pair of millworkers' cottages.

the fell, moor or river bed. The pioneer student of the history of vernacular buildings in the Dales, James Walton, quotes from an account paid by the vicar of Kirkby Malham in 1454 which mentioned '. . . for drink given to the carpenters and for basyng the said houses, that is to say, for laying great stones under the foot of the Crokk, 4d.' While it is clear from the account that the main walls of the house were of wattle, and the roof of thatch or turf or ling, the cruck blades plainly rested on a base of stone. Hardly any thatched, wattle-walled and obviously cruck-framed farmsteads or barns survive in the Dales, but there are a good number where the cruck frame survives, encased within stone walls and invisible from the outside. Not many Pennine long-houses remain in occupation. In the late eighteenth and nineteenth centuries, many were succeeded by two-storey stone farmsteads with symmetrical, crudely Classical façades. However, the long-house often persists, humbled and converted into a barn.

Just as gritstone walls succeeded wattle in the course of the sixteenth, seventeenth and eighteenth centuries, so roofs composed of great sandstone slabs, of the type which had been used in some of the greater Norman halls, were diffused downwards through the social grades to replace the thatched roofs of lowlier farmsteads. These slabs are known as 'thackstones'. They were quarried at places like Hawes, Askrigg, Carberry and Low Reeth, from beds of easily-split, or 'fissile', strata interbedded with the local limestone, or on Fountains Fell, where fissile sandstones were found at the basal levels of the Millstone Grit. Split into thin sheets, the thackstones were carefully graded according to size into up to eighteen classes, each class having its dialect name: 'Scant Fairwells' just ten-and-a-half inches long were the shortest, 'Thirteens' measuring thirty-six inches were the longest, while 'Long Skirtchens', 'Batchelors' and 'Wivetts' were among the intervening grades.

Oak pegs driven through holes chipped in the slates and into the oak rafters secured the thackstones to the roof, and the courses were graded, with the lengths of slates decreasing towards the ridge. The lowest course rested upon a second under-eaves course of slates which was mortared to the wall, the gap between the two layers being packed with small wedging stones for a snug and stable fit. Being much heavier than the stone tile roofs of the Midlands limestone belt, these Pennine roofs have a minimal pitch of around thirty-eight degrees to reduce the amount of stone carried on the rafters. Although thackstone roofs were neither wind nor blizzard proof, the gaps in them were seldom sealed with mortar; sphagnum moss was often packed between the cracks. Sadly, these roofs, so characteristic of the Pennine landscape, have proved difficult to maintain, and now many of them have been replaced by incongruous red pantiles or thin blue sheets of Welsh slate.

Our heritage of stone cottages and farmsteads of the seventeenth, eighteenth and early nineteenth centuries belongs to the late autumn and early winter of the stone-building era. Some houses in stone or reconstituted stone (which is sometimes quite ghastly) are still being built in areas where planning regulations seek harmony between the new and the old. Although planning regulations in some areas of distinguished vernacular stone building require the use of stone, the 'indigenous' material may be costly or no longer obtainable. In the Yorkshire Dales, for example, Millstone Grit is now very expensive. Many new stone houses, like my own, employ a similar coarse gritstone quarried near Bradford, and which is often recycled from the Coal Measures, which is often recycled from demolished industrial buildings. The stone cottages of the Cotswolds, Cornwall, Wales and the other stony provinces were built by village roughmasons, often assisted by their owners or tenants and their neighbours. The people who caused them to be built told the mason exactly what they wanted and where, and then craftsmanship fulfilled the task. Architects were not thought of. Yet there is no question that the old dwellings belong more comfortably in their setting, and that these are what the summer trippers flock to admire. Even

A distinctive vernacular building tradition in stone is represented by the so-called 'black houses' of the Scottish Highlands and Islands. In many examples, the thick stone walls were built hollow and packed with a draught-proofing filling of earth. This house is at Colbost on Skye and its reconstructed interior is open to the public.

The vernacular tradition in stone walling and roofing displayed at the old quarrying village of Corfe in Dorset.

when old stone farmsteads are vandalized by glazed doors and ugly modern windows, festooned with aerials and hung with telephone wires, much of their original charm may still shine through.

Two stone building forms spanned the passing ages. Being essential to the economic life of the community, bridges and field walls were less susceptible to fashions and social changes than ecclesiastical and domestic buildings. The farmer must delimit his holding, keep the cows from the corn, and divide his farm into functional packages. In the creation of field boundaries and barriers, the choice was largely between the hedge and the stone wall. Other things being equal, either might be used; and hedges and walls are often seen side by side in the Cotswolds, Dales and beneath the Lakeland fells. In its initial stages, at least, the blackthorn and hawthorn hedge is cheaper and more rapidly extended across the landscape. It will regenerate itself indefinitely, while acquiring one new hedgerow species of tree or shrub around every century in every thirty-yard section. However, in order to maintain a trim, vigorous stock-proof screen, time-consuming annual maintenance is necessary. The wall requires less maintenance but more building effort and abundant supplies of cheap stone rubble. Fortunately, the chilly windswept uplands, where hawthorn will not succeed, tend to be the areas which abound in scree, moorstone or stream-bed boulders.

Field walls vary considerably from one region of Britain to another, and each type is expressive of both local geology and the purposes which it is built to serve. Changes may also exist within any local network of walls. The largest walls are normally those which were built to define the boundaries of estates, and while the estates may be fragmented, their walls endure for centuries. In the Dales to the west of Ripon, for example, the thick walls built on earthen banks run for mile after mile over moor, fell and valley to trace out the long-defunct holdings of Fountains Abbey. Walls which run straight and true, carving the landscape into rectangular parcels, preserve the allocations of the eighteenth- and nineteenth-century Parliamentary Enclosure commissioners, while those which curve and ramble are

A modern professional drystone waller at work in the Yorkshire Dales.

generally older, and signify agreements reached between neighbouring landowners in the course of the centuries preceding the official Enclosure movement. Below Malham Cove, in parts of Wharfedale and around Castleton in Derbyshire, one can see walls defining ribbon-like fields which are directly descended from medieval field strips. There is more to a wall than may initially meet the eye.

We have already seen that the stone wall has a pedigree extending into prehistoric times. During the post-medieval era, some walls were built by specialists known as 'dykers', others by village roughmasons while they were not engaged on domestic projects, and others still by the farmer or his helpers. Each region of the country developed the type of wall which was most in keeping with its assets and its needs. In the Pennine uplands and the Scottish mountains, walls must be sufficiently substantial to withstand winter gales and stable enough to cope with the extra gravitational problems posed by an upslope course. In parts of Scotland, a type of wall known as a 'Galloway dyke' evolved; the lower forty inches are built solidly in a double thickness of stone, while the topmost twenty inches consists of a single thickness of loosely-spaced stones. It is said that the athletic mountain sheep will not breach or scramble a wall so long as it can glimpse what lies on the other side – hence the gaps in the upper levels. In more northerly places, notably the north-east, the agricultural Improvers of the eighteenth and nineteenth centuries had the boulder litter from their fields gathered and contained in massive 'consumption dykes'.

Parliamentary Enclosures of the late eighteenth and early nineteenth centuries provided work for wallers as well as hedgers and nurserymen. These gritstone walls are on the plateau above Nidderdale.

In the Pennines, the walls closely reflect the geology beneath, being of Carboniferous limestone – frost-shattered into angular chunks – or of the darker and generally less jagged forms of Millstone Grit boulders gathered from fields, moors or stream beds. In the Cotswolds, where the finest networks of drystone walling are displayed, the walls are lower – the landscape is more plump and mellow, and so are the sheep. Thus, the walls here are seldom more than three feet in height and need only be built to a single thickness. However, in order to discourage attempts by the sheep to scale them, an irregular crest of narrow vertical slabs or 'combers' is often mortared to the horizontal course beneath. The limestones of the Cotswolds and the other regions of the oolite belt are easily split by hammers, or by the action of frost, to form the narrow slabs which are ideal for walling purposes – in this respect, as in all others, they are beloved by roughmason and freemason alike.

Among the many other variations upon the theme of drystone walling, two are quite distinctive. I have already sung the praises of the remarkable Caithness flagstone of Orkney, and, in odd corners of the islands, one may still see the simplest of all walls, consisting of vertical flagstone slabs set in line like the scales on the back of a stegosaurus. In the slate districts of west Wales, the slates often appear in walls built to a herringbone pattern. Around an inch in thickness and roughly shaped to a rectangular form, the slates are laid in courses which alternately slope around thirty degrees to either side of the vertical. The resulting herringbone effect is not intended for decoration but for strength, as each course is tightly wedged upon the one beneath; and as humus accumulates in the fissures of the courses, the wall becomes bonded by an overgrowth of grasses.

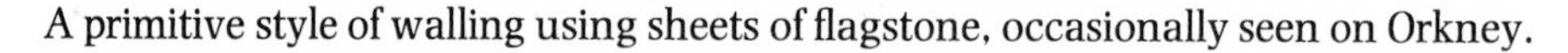
A primitive style of walling using sheets of flagstone, occasionally seen on Orkney.

This style of walling with a top row of 'combers' bedded in mortar was favoured in the sheep country of the Cotswolds.

While British stone walls vary according to local circumstances, certain qualities were generally sought. Though it was considered wrong to concentrate too much stone in the lower courses, a slight batter or upward taper was desired. Stones were expertly laid to achieve close contacts, but a certain amount of spacing improves air circulation and reduces damage from frost shattering, caused by the expansion of water freezing in airtight pockets. In the case of the larger walls, which were built of a double thickness of stone, a filling, or 'hearting', of small stones was sandwiched between the outer courses, while large 'through stones' were set at regular intervals, running right through the wall, to bond the opposing courses.

The makers of drystone walls relied upon their craftsmanship and experience rather than upon special tools, and a good dyker might lay around twenty feet of wall in the course of a day. If he had done his work well, the wall might last for a century or more without repair. A length of string and a frame of sticks outlining a cross-section of wall were all that was needed to maintain a constant height and profile; leather hand pads, a heavy hammer, and perhaps a small stone pick completed the tool-kit.

The survival of our legacy of walls is threatened by the declining number of competent dykers, and by heavy lorries which, in using country roads, shake and brush the roadside walls and render them more liable to collapse. Fortunately, most walls are associated with areas of upland livestock farming, and so the destruction cannot compare with that now being caused by the grubbing out of lowland hedgerows for the creation of prairie-like arable fields. (This is the greatest single threat to our beleaguered countryside. Without swift and forceful action, the lowland scene will be reduced to a featureless desert in the course of this generation.)

The first British bridges of real substance were built by the Romans. None of these survive, but in a number of places the remains of their stone abutments have been found. Although their virtues go largely unsung, the fine bridges created by medieval masons display a craftsmanship exceeded only by that shown in the greater churches. Many have been transformed by road-widening operations, but a considerable number survive almost intact, including the quite well-known examples at Chester, at St Ives in Cambridgeshire, and at Bradford-upon-Avon in Wiltshire, the latter two being complete with bridge chapels. In addition, there are scores of less familiar survivors, like the attractive examples at Ripley in Yorkshire, Pershore in Worcestershire, and the series of grand old bridges which link Devon and Cornwall across the River Tamar. The evolutionary stages in the development of stone-bridge designs are more subtle than those displayed in buildings, but the medieval bridge can often be recognized by its arches – which may be pointed in the Gothic mode – by the pointed cutwaters which divert the oncoming waters from the piers, or by the ribbed stonework on the undersides of the arches, which is designed to provide strength while economizing on materials. These, however, are not hard and fast rules, and stone-bridge designs had changed little by the eighteenth century, the essential problems of bridge-building having been mastered in the Middle Ages.

In the post-medieval centuries, the volume of traffic steadily increased, and more and more perishable and vulnerable wooden bridges were replaced by bridges of stone. At the same time, the growth of the pack-horse and droving trade stimulated the construction of scores of single-arched and narrow-waisted pack-horse bridges to ease movement along rocky upland tracks. Few of these charming little bridges are as old as they seem, a high proportion dating from the eighteenth century. Far more primitive in their appearance, but not necessarily even medieval in age, are the clapper bridges, consisting of great horizontal stone slabs set upon piers of piled boulders. Though these are most common on Dartmoor, other examples are quite widely dispersed. While many a wooden bridge yielded to a stone successor, in the course of the nineteenth century, new bridges were often built of iron girders; and for many more recent years, the concrete bridge has been the norm.

A typical pack-horse bridge in the Lake District. Such bridges date mainly from the eighteenth and nineteenth centuries and were often built by the local farmers. This one served the route across Sty Head Pass.

It is perhaps appropriate that one branch of the craft of masonry which continued to flourish during the declining years of the age of stone building was that of tombstone carving. While the medieval lord and his lady might lie at rest in a tomb chest beneath their full-sized effigies, the lesser folk of the parish would be buried outside the church in shallow graves around a single churchyard cross. Gradually, fashions changed, and in the post-medieval period tombstones began to appear in the churchyard. Seventeenth-century tombstones are rarely seen, but when they can be found they are often attractive – in a morbid sort of way. While church monuments to the mighty were tending to become florid and maudlin, the tombstones of the middle classes dealt more directly with the realities of death. Skulls and skeletons are the favoured themes. Orwell church in Cambridgeshire contains a splendidly stark tombstone and grave-slab dating from 1632. Eighteenth-century tombstones, generally carved from good limestone, are much more common and almost invariably elegant. Sadly, they are often to be seen stacked against churchyard walls, displaced by

In the industrial era, limestone has proved to be vulnerable to attack by acid rain, and unless there is effective legislation then much of the heritage will be destroyed. These photographs show the effects of erosion at York Minster (left) and on the façade at Wells Cathedral (right), subject of an extensive recent restoration.

the more flamboyant and melodramatic creations of the nineteenth century. The development of the railways gave monumental masons access to a wide array of brightly-coloured granites and marbles. Even in death, British individuality found expression in tombstones and monuments, so that the churchyard became a discordant symphony of clashing shapes and hues. Many churchyards, particularly those of the Jurassic limestone belt, like Wadenhoe and Brigstock in Northamptonshire, preserve assemblages of eighteenth-century stones. Restful, mellow and harmonious, the older stones complement the spirit of the older church nearby.

We have now reached the deep winter of the age of stone building, and it is almost inconceivable that a new spring will ever dawn. However, it would be wrong to assume that the decline in stone building since the close of the eighteenth century was precipitous and unrelieved. Mixed in with the maelstrom of building operations which accompanied the later stages of the Industrial Revolution was a programme of constructing churches and civic buildings in stone which, in terms of activity if not always in the quality of output, must have equalled that of a medieval century. Assured and dynamic as they were, the Victorians did not develop a single coherent building style of their own, but tended to borrow, jackdaw-like, from older traditions, particularly coloured by a romantic love of the various Gothic designs.

The greatest source of employment for modern masons is restoration work. These photographs, from the workshops of Rattee and Kett in Cambridge, show a severely-eroded original stone lion's head (left) and its replacement (right).

Much that resulted was undistinguished or bizarre, but many a young and spreading industrial centre received a neo-Gothic church that was well proportioned and finely wrought.

In their alterations to cathedrals and country mansions, the Victorians did a great deal that modern arbiters of taste and conservationists deplore, but we should also remember the scores of churches and monuments which were saved by their timely restoration after centuries of neglect and decay. The more closely one looks at the carvings upon the exteriors of medieval churches, the more one becomes aware of the amazing scope of Victorian restoration campaigns, for so much that passes at first glance for medieval work proves to belong to the last century. Despite the heavy-handedness of much of their work, we owe a great debt to the Victorian restorers; but the tradition of masonry could not progress through the reproduction or parodying of time-worn themes. The tradition of masonry was going nowhere.

As these chapters have shown, stone buildings express the conditions of their age. Ours is an age of mass building, of tight time schedules and cost effectiveness, of architects, auditors and concrete. The most impressive sight the average medieval peasant was ever likely to see was a new, stone-built church or cathedral, 'Glowying as the sunne'. Buildings such as these provided the only glamour and wonder that such people might know and we can be sure that the people stood spellbound in their shadows. The great

Stone in the masons' yard destined for restoration work at the adjacent York Minster.

In certain areas that are noted for their stone building traditions, the planning regulations require the use of stone for new dwellings. Often, as here in the Yorkshire Dales, the stone employed is re-cycled from demolished industrial buildings.

Modern restoration work at Byland Abbey in Yorkshire.

medieval churches and castles, as well as the later stately homes, stone cottages and farmsteads, still appeal in their different ways. The best of them even inspire. Work over, we flock to see them and try to breathe the air of distant ages. We will never have the resources to build like this again. The best that we can do is to try to preserve the heritage – and we do not always do this very well.

Unswervingly resolute once their anger has been aroused, in matters of conservation the British tend to fall victims to their passive national character. While 'muddling along' has become a way of life, it is generally assumed that governments can be entrusted to care for the heritage, and that Britain still leads the world in fields such as preservation and restoration. Buildings and archaeological sites which Americans would cherish are demolished and bulldozed, while the restoration works achieved in some much poorer Eastern European states put the British efforts into an extremely unflattering perspective. In some cases, the national or local authorities protect or restore a building of merit, and in others they do not. The real encouragement comes not from the erratic achievements of government, but from the efforts of private individuals, local amenity societies and conservation groups. There are scores of little communities which have given the time, energy and donations necessary to raise the many thousands of pounds needed to restore a parish church, while several of our finest cathedrals would be crumbling and unsafe without the voluntary contributions from visitors, tourists and appeal funds. However, until conservation becomes a political issue, as it did in parts of the USA during the 1970s, destruction will continue to outstrip preservation.

Few of us would wish to return to the ages of mass poverty, exploitation and injustice to which most of our finest architectural creations belong. At the same time, our modern materialistic society seems incapable of producing buildings and monuments which are ever much more than functional, or are capable of exciting a popular enthusiasm. It would be wrong to cast governments, councils and architects as the sole scapegoats. The architect Sir Hugh Casson writes: 'Buildings speak louder than words, and no amount of protestation from my profession about "forces beyond our control" or the nature of the acquisitive and apathetic society we serve, can cancel out the faceless banality of what we see around us.' When we have had more time to explore the limits of materialism, perhaps we will become less acquisitive and less apathetic, and then we may discover better ways of blending a rewarding enthusiasm for the past with our hopes for the future.

Further Reading

BASIC GEOLOGY

P. H. Armstrong *Discovering Geology* (Shire, 1974)
D. Dineley *Rocks* (Collins Countryside Series, Collins, 1975)
W. R. Hamilton, A. R. Woolley and A. C. Bishop *The Hamlyn Guide to Minerals, Rocks and Fossils* (Hamlyn, 1974)
E. H. Shackleton *Geological Excursions in Lakeland* (Dalesman, 1975)
R. M. Wood *On the Rocks* (BBC, 1978)

STONE IN THE LANDSCAPE

A. D. Trueman *Geology and Scenery in England and Wales* (Penguin, 1971)
J. A. Steers *The Coast of England and Wales in Pictures* (CUP, 2nd edn, 1960)
R. Millward and A. Robinson *Landscapes of Britain* (David & Charles, 1977)
The various regional volumes in the Methuen *Geomorphology of the British Isles* series ed. E. H. Brown and K. Clayton.

STONE AND PREHISTORIC MAN

A. Burl *The Stone Circles of the British Isles* (Yale, 1976)
K. P. Oakley *Man the Tool-maker* (British Museum, 1972)
W. Watson *Flint Implements* (British Museum, 3rd edn, 1965)
R. Feachem *Guide to Prehistoric Scotland* (Batsford, 2nd edn, 1977)
C. Hodder *Wales: An Archaeological Guide* (Faber, 1974)
S. P. O'Riordain *Antiquities of the Irish Countryside* (Methuen 5th edn, 1979)
N. Thomas *Guide to Prehistoric England* (Batsford, 2nd edn, 1976)

QUARRYING AND MASONRY

M. A. Aston *Stonesfield Slate* (Oxfordshire CC Museums Service, 1974)

J. A. Best, S. Parker and C. M. Prickett *The Lincolnshire Limestone* (1981–obtainable from Nene College, Northamptonshire)

A. Clifton-Taylor *The Cathedrals of England* (Thames & Hudson, 1967)

A. Clifton-Taylor and A. S. Ireson *English Stone Building* (Gollancz, 1983)

D. Knoop and G. P. Jones *The Medieval Mason* (Manchester UP, 3rd edn, 1967)

L. and J. Llaing *A Guide to Dark Age Remains in Britain* (Constable, 1979)

J. and J. Penoyre *Houses in the Landscape* (Faber, 1978)

L. F. Salzman *Building in England* (OUP, 1952)

M. Wood *The English Mediaeval House* (Ferndale, 1981)

Index

Page numbers in *italic* refer to the illustrations

Index